Direct Leadership

Keeping it Simple Stupid

MICHAEL F. MORTON

ISBN: 978-0-9995832-8-9

DEDICATION

This book is dedicated to those who lead, and those who follow.

CONTENTS

ACKNOWLEDGMENTS

There are and have been so many leaders and mentors in my life. Some I hold in the highest esteem, monuments to how I live my life. Others I do not, but I learned as much from them, as I did the mentors in my military and civilian career.

I wrote this book using the time-honored KISS method. If you are not familiar with this method, it stands for Keep It Simple Stupid. I did not use fancy graduate level vocabulary or complex algorithms to write this book. It is designed to be user friendly and doctrinally sound.

1 WHAT IS LEADERSHIP?

You've likely had a manager or two in your career that you weren't very fond of. And on the flip side, you've probably worked with or for someone who you looked up to and admired a lot. Managers and leaders. Not every manager is good at leadership, and not every great leader is good at management. There are arguments about if these two terms are the same or different. What most research has come to find is that a manager is more about your position, and leadership is more about who you are. You can be a manager and a leader, or you can be one or the other. If you're a current or aspiring manager, it's really important to focus on good leadership. There are some differences between these two titles, but it can be a huge asset to understand these differences and how you can work to develop the characteristics of both.

Managers are generally responsible for the day-to-day operations of an operation. They have people who report to them, they oversee their work, and help make all the projects run smoothly. They manage systems confidently and help their workers feel understood and meaningful to the organization.

Leadership is all about building a vision for people to follow. They inspire and direct with authority and intelligence. They help people feel confident and excited to follow their lead.

Leadership and management may seem to be different, but they share many qualities. If a manager is a good leader, their workers will be willing to follow their lead, they will trust them, and they will have more positive work experience.

Leadership has nothing to do with seniority or one's position in the hierarchy of an organization. Too many talk about an organization's leadership referring to the senior most executives in the organization. They are just that, senior executives. Leadership doesn't automatically happen when you reach a certain pay grade. Hopefully you find it there, but there are no guarantees.

Leadership has nothing to do with titles. Like the point above, just because you have a C-level (Chief ----) title, doesn't automatically make you a "leader." In all the leadership courses I teach, I stress the fact that you don't need a title to lead. In fact, you can be a leader in your place of worship, your neighborhood, in your family, all without having a title.

Leadership has nothing to do with personal attributes. Say the word "leader" and most people think of a domineering, take-charge charismatic individual. We often think of icons from history like General Patton or President Lincoln. But leadership is not an adjective. We don't need extroverted charismatic traits to practice leadership. And those with charisma don't automatically lead.

Leadership isn't management. This is the big one. Leadership and management are not synonymous. You have 15 people in your downline and P&L responsibility? Good for you, hopefully you are a good manager. Good management is needed. Managers need to plan, measure, monitor, coordinate, solve, hire, fire, and so many other things. Typically, managers manage things. Leaders lead people.

My definition of leadership: Leadership is an Art, a process of inspiring and influencing people to do things. Leadership is not a sometime thing, it's an all-time thing. You don't lead once in a while; you must be a leader at all the times.

Leadership is fluid, like water. Water can flow, it can crash, it can penetrate any surface. Put into a bottle, it becomes the bottle, put into people, and they will flourish. "Be like water". (Bruce Lee)

Types of Leaders/Leadership

Transactional Leadership

This leadership style starts with the idea that team members agree to obey their leader when they accept a job. The "transaction" usually involves the organization paying team members in return for their effort and compliance. The leader has a right to "punish" team members if their work doesn't meet an appropriate standard.

Although this might sound controlling and paternalistic, transactional leadership offers some benefits. For one, this leadership style clarifies everyone's roles and responsibilities. Another benefit is that, because transactional leadership judges team members on performance, people who are ambitious or who are motivated by external rewards – including compensation – often thrive.

The downside of this leadership style is that team members can do little to improve their job satisfaction. It can feel stifling, and it can lead to high staff turnover.

Transactional leadership is really a type of management, not a true leadership style, because the focus is on short-term tasks. It has serious limitations for knowledge-based or creative work. However, it can be effective in other situations

Autocratic Leadership

Autocratic leadership is an extreme form of transactional leadership, where leaders have complete power over their people. Staff and team members have little opportunity to make suggestions, even if these would be in the team's or the organization's best interest.

The benefit of autocratic leadership is that it's incredibly efficient. Decisions are made quickly, and work gets done.

The downside is that most people resent being treated this way. Therefore, autocratic leadership often leads to high levels of absenteeism and high staff turnover. However, the style can be effective for some routine and unskilled jobs: in these situations, the advantages of control may outweigh the disadvantages.

Autocratic leadership is often best used in crises when decisions must be made quickly and without dissent.

Bureaucratic Leadership

Bureaucratic leaders work "by the book." They follow rules rigorously and ensure that their people follow procedures precisely.

This is an appropriate leadership style for work involving serious safety risks (such as working with machinery, with toxic substances, or at dangerous heights) or where large sums of money are involved. Bureaucratic leadership is also useful in organizations where employees do routine tasks (as in manufacturing).

The downside of this leadership style is that it's ineffective in teams and organizations that rely on flexibility, creativity, or innovation.

Charismatic Leadership

A charismatic leadership style can resemble transformational leadership because these leaders inspire enthusiasm in their teams and are energetic in motivating others to move forward. This excitement and commitment from teams is an enormous benefit.

The difference between charismatic leaders and transformational leaders lies in their intention. Transformational leaders want to transform their teams and organizations. Charismatic leaders are often focused on themselves and may not want to change anything.

The downside to charismatic leaders is that they tend believe more in themselves than in their teams. This can create the risk that a project or even an entire organization might collapse if the leader leaves. A charismatic leader might believe that she can do no wrong, even when others are warning her about the path she's on; this feeling of invincibility can ruin a team or an organization.

Democratic/Participative Leadership

Democratic leaders make the final decisions, but they include team members in the decision-making process. They encourage creativity, and team members are often highly engaged in projects and decisions.

There are many benefits of democratic leadership. Team members tend to have high job satisfaction and are productive because they're more involved in decisions. This style also helps develop people's skills. Team members feel in control of their destiny, so they're motivated to work hard by more than just a financial reward. This approach can slow decision-making.

The downside of democratic leadership is that it can often hinder situations where speed or efficiency is essential. For instance, during a crisis, a team can waste valuable time gathering people's input. Another downside is that some team members might not have the knowledge or expertise to provide high quality input.

Laissez-Faire Leadership

Laissez-Faire means "leave it be," and it describes leaders who allow their people to work on their own. Laissez-faire leaders may give their teams' complete freedom to do their work and set their own deadlines. They provide team support with resources and advice, if needed, but otherwise don't get involved.

This leadership style can be effective if the leader monitors performance and gives feedback to team members regularly. It is most likely to be effective when individual team members are experienced, skilled, self-starters.

The main benefit of laissez-faire leadership is that giving team members so much autonomy can lead to high job satisfaction and increased productivity.

The downside is that it can be damaging if team members don't manage their time well or if they don't have the knowledge, skills, or motivation to do their work effectively.

Task-Oriented Leadership

Task-oriented leaders focus only on getting the job done and can be autocratic. They actively define the work and the roles required, put structures in place, and plan, organize, and monitor work. These leaders also perform other key tasks, such as creating and maintaining standards for performance.

The benefit of task-oriented leadership is that it ensures that deadlines are met, and it's especially useful for team members who don't manage their time well.

However, because task-oriented leaders don't tend to think much about their team's well-being, this approach can suffer many of the flaws of autocratic leadership, including causing motivation and retention problems.

People-Oriented/Relations-Oriented Leadership

With people-oriented leadership, leaders are totally focused on organizing, supporting, and developing the people on their teams. This is a participatory style and tends to encourage good teamwork and creative collaboration. This is the opposite of task-oriented leadership.

People-oriented leaders treat everyone on the team equally. They're friendly and approachable, they pay attention to the welfare of everyone in the group, and they make themselves available whenever team members need help or advice.

The benefit of this leadership style is that people-oriented leaders create teams that everyone wants to be part of. Team members are often more productive and willing to take risks because they know that the leader will provide support if they need it.

The downside is that some leaders can take this approach too far; they may put the development of their team above tasks or project directives.

Servant Leadership

A Leader often not formally recognized as such. When someone at any level within an organization leads simply by meeting the needs of the team, (situational Leader). They have high integrity and lead with generosity.

Servant leaders often "lead from behind," preferring to stay out of the limelight and letting their team accept recognition for their hard work.

People who practice servant leadership can find themselves left behind by leaders using other leadership styles. This leadership style also takes time to apply correctly: it's ill-suited in situations where you must make quick decisions or meet tight deadlines.

Transformational Leadership

Transformational leaders are inspiring because they expect the best from everyone on their team as well as themselves. This leads to high productivity and engagement from everyone in their team.

The downside of transformational leadership is that while the leader's enthusiasm is passed onto the team, he or she can need to be supported by "detailed people."

Are leaders born or made?

Leadership is a learned behavior that becomes unconscious and automatic over time. For example, leaders can make several important decisions about an issue in the time it takes others to understand the question. Many people wonder how leaders know how to make the best decisions, often under immense pressure. The process of making these decisions comes from an accumulation of experiences and encounters with a multitude of different circumstances, personality types and unforeseen failures. More so, the decision-making process is an acute understanding of being familiar with the cause and effect of behavioral and circumstantial patterns; knowing the intelligence and interconnection points of the variables involved in these patterns allows a leader to confidently make decisions and project the probability of their desired outcomes.

The most successful leaders are instinctual decision makers. Having done it so many times throughout their careers, they become immune to the pressure associated with decision making and extremely intuitive about the process of making the most strategic and best decisions. This is why most senior executives will tell you they depend strongly upon their "gut-feel" when making difficult decisions at a moment's notice.

Beyond decision making, successful leadership across all areas becomes learned and instinctual over a period. Successful leaders have learned the mastery of anticipating business patterns, finding opportunities in pressure situations, serving the people they lead and overcoming hardships. No wonder the best CEOs are paid so much money. In 2011, salaries for the 200 top-paid CEOs rose 5 percent to a median $14.5 million per year, according to a study by compensation-data company Equilar for The New York Times. (NYT, 2011)

2 WHAT MAKES A LEADER

In today's fast-paced, technology-driven world, strong leadership is more important than ever. But what makes a good leader? And how does strong leadership impact the workplace? In this section I will cover several areas of what makes a leader, including Principles, Traits, Qualities, and Duties. The information in this chapter was taken from U.S. Army Leadership manuals FM 22-100 Army Leadership, Be, Know, Do; and FM 6-22 Leader Development

Leadership Principles

1. Know Yourself and Seek Self Improvement...Develop a plan to keep your strengths and improve on your weaknesses.

2. Be Technically Proficient...Not only do we know our duties and responsibilities, but we also know all those of our team members, and we look to our leaders and concern ourselves with learning their duties and responsibilities.

3. Seek Responsibility and Take Responsibility for Your Actions...We are not satisfied with performing just our duties to the best of our abilities, we look to grow and seek further challenges, and always, when in charge, accept the consequences of our decisions, absorb the negative and pass on the praises.

4. Make Sound and Timely Decisions...Leaders must be able to reason under the most critical condition and decide quickly what action to take.

5. Set the Example...No aspect of leadership is more powerful. Our personal example affects people more than any amount of instruction or

form of discipline. You are the role model.

6. Know Your Personnel and Look Out for Their Well Being...Leaders must know and understand those being led. When individuals trust you, they will willingly work to help accomplish any mission.

7. Keep Your Staff Informed...Our team members expect us to keep them informed, and where and when possible, to explain the reasons behind requirements and decisions. Information encourages initiative, improves teamwork, and enhances morale.

8. Develop A Sense of Responsibility In Your Followers...The members of a team will feel a sense of pride and responsibility when they successfully accomplish a new task given to them. When we delegate responsibility to our followers, we are indicating that we trust them.

9. Ensure Each Task is Understood, Supervised and Accomplished...Team members must know the standard. Supervising lets us know the task is understood and lets our team members know we care about mission accomplishment and about them.
insure mission accomplishment.

10. Build A Team...Leaders develop a team spirit that motivates team members to work with confidence and competence. Because mission accomplishment is based on teamwork, it is evident the better the team, the better the team will perform the task

11. Develop Your Team In Accordance With Its Capabilities...A leader must use sound judgment when developing the team. Failure is not an option. By employing the team properly, we insure mission accomplishment.

Leadership Traits

JUSTICE

Definition: Justice is defined as the practice of being fair and consistent. A just person considers each side of a situation and bases rewards or punishments on merit.

Suggestions for Improvement: Be honest with yourself about why you make a particular decision. Avoid favoritism. Try to be always fair and treat all things and people in an equal manner.

JUDGMENT

Definition: Judgment is your ability to think about things clearly, calmly, and in an orderly fashion so that you can make good decisions.

Suggestions for Improvement: You can improve your judgment if you avoid making rash decisions. Approach problems with a commonsense attitude.

TACT

Definition: Tact means that you can deal with people in a manner that will maintain good relations and avoid problems. It means that you are polite, calm, and firm.

Suggestions for Improvement: Begin to develop your tact by trying to be always courteous and cheerful. Treat others as you would like to be treated.

DEPENDABILITY

Definition: Dependability means that you can be relied upon to perform your duties properly. It means that you can be trusted to complete a job. It is the willing and voluntary support of the policies and orders of the chain of command. Dependability also means consistently putting forth your best effort to achieve the highest standards of performance.

Suggestions for Improvement: You can increase your dependability by forming the habit of being where you're supposed to be on time, by not making excuses and by carrying out every task to the best of your ability regardless of whether you like it or agree with it.

INITIATIVE

Definition: Initiative is acting even though you haven't been given orders. It means meeting new and unexpected situations with prompt action. It includes using resourcefulness to get something done without the normal material or methods being available to you.

Suggestions for Improvement: To improve your initiative, work on staying mentally and physically alert. Be aware of things that need to be done and then to do them without having to be told.

DECISIVENESS

Definition: Decisiveness means that you can make good decisions without delay. Get all the facts and weight them against each other. By acting calmly and quickly, you should arrive at a sound decision. You announce your decisions in a clear, firm, professional manner.

Suggestions for Improvement: Practice being positive in your actions instead of acting half-heartedly or changing your mind on an issue.

INTEGRITY

Definition: Integrity means that you are honest and truthful in what you say or do. You put honesty, sense of duty, and sound moral principles above all else.

Suggestions for Improvement: Be always absolutely honest and truthful. Stand up for what you believe to be right.

ENTHUSIASM

Definition: Enthusiasm is defined as a sincere interest and exuberance in the performance of your duties. If you are enthusiastic, you are optimistic, cheerful, and willing to accept the challenges.

Suggestions for Improvement: Understanding and belief in your mission will add to your enthusiasm for your job. Try to understand why even uninteresting jobs must be done.

BEARING

Definition: Bearing is the way you conduct and carry yourself. Your manner should reflect alertness, competence, confidence, and control.

Suggestions for Improvement: To develop bearing, you should hold yourself to the highest standards of personal conduct. Never be content with meeting only the minimum requirements.

UNSELFISHNESS

Definition: Unselfishness means that you avoid making yourself comfortable at the expense of others. Be considerate of others. Give credit to those who deserve it.

Suggestions for Improvement: Avoid using your position or rank for personal gain, safety, or pleasure at the expensive of others. Be considerate of others.

COURAGE

Definition: Courage is what allows you to remain calm while recognizing fear. Moral courage means having the inner strength to stand up for what is right and to accept blame when something is your fault. Physical courage means that you can continue to function effectively when there is physical danger present.

Suggestions for Improvement: You can begin to control fear by practicing self-discipline and calmness. If you fear doing certain things required in your daily life, force yourself to do them until you can control your reaction.

KNOWLEDGE

Definition: Knowledge is the understanding of a science or art. Knowledge means that you have acquired information and that you understand people. Your knowledge should be broad, and in addition to knowing your job, you should know your unit's policies and keep up with current events.

Suggestions for Improvement: Suggestions for Improvement: Increase your knowledge by remaining alert. Listen, observe, and find out about things you don't understand.

LOYALTY

Definition: Loyalty means that you are devoted to your country, your organization, department, and to your seniors, peers, and subordinates.

Suggestions for Improvement: To improve your loyalty you should show your loyalty by never discussing the problems of the department with outsiders. Never talk about seniors unfavorably in front of your subordinates. Once a decision is made and the order is given to execute it, carry out that order willingly as if it were your own.

ENDURANCE

Definition: Endurance is the mental and physical stamina that is measured by your ability to withstand pain, fatigue, stress, and hardship. For example, enduring pain during a conditioning march in order to improve stamina is crucial in the development of leadership.

Suggestions for Improvement: Develop your endurance by. Finish every task to the best of your ability by forcing yourself to continue when you are physically tired, and your mind is sluggish.

Leadership Qualities

Take a minute to look at yourself as a leader. Do you have the qualities to make a valuable leader to your group? Review the leadership qualities and put yourself to the test. After you have looked at yourself as a leader, answer the following questions

Attributes . . .

- Do I view problems as opportunities?
- Am I a priority setter?
- Am I customer focused?
- Am I courageous?
- Am I a critical and creative thinker?
- What is my tolerance for ambiguity?
- Am I positive attitude towards change?
- Am I committed to innovations that are best for children?

Skills . . .

- Do I debate, clarify, and enunciate my values and beliefs?
- Can I fuel, inspire, and guard the shared vision?
- Can I communicate the strategic plan at all levels?
- Do I recognize the problems inherent to the planning process?

- Do I ask the big picture questions and "what if"?
- Can I support the school staff through the change process?
- Do I encourage dreaming and thinking the unthinkable?
- Can I align the budget, planning, policies and instructional programs with the district goals and vision?
- Do I engage in goal setting?
- Can I develop and implement action plans?
- Do I practice and plan conscious abandonment?
- Do I transfer the strategic planning process to planning?

Knowledge . . .

- Do I know board and superintendent roles and responsibilities in planning and implementing plans?
- Do I know the strategic planning process, short- and long-term planning tools?
- Do I know the board and district vision, beliefs, and mission?
- Do I know the relationship of the budget to district planning?
- Do I know local, state, and national factors that affect education?
- Do I know the best practices and research on improving academic achievement?
- Do I know the process of change and paradigm shifts?
- Do I know the strategies to involve and communicate with the community?

Leadership Behaviors

<u>Establish a clear direction</u>

a) Envision the future

b) Focused attention

c) Articulate values

d) Enlist others

<u>Communicate, communicate, communicate</u>

a) Provide meaning

b) Utilize all forms

c) "Fire in belly"

Position the Organization

a) Build trust (encourage the heart)

b) Reliability and constancy

c) Plan small wins

d) Recognize contributions

e) Celebrate accomplishments

Development of self

a) Set the example (walk like you talk)

b) Positive self-regard

c) Focus on winning

d) Search for opportunities

e) Experiment and take risks

Empower followers/members

a) Establish clear buy-in

b) Foster collaboration

c) Develop a supportive environment

Leadership Duties:

Specific: duties are those related to jobs and positions. This includes regulations, policies, and procedures. Example: Conduct training courses

Directed: duties are not specified as part of a job position. A supervisor gives them orally or in writing. Examples: Conducting problem solving huddles

Implied duties: often support specified duties, but in some cases, they may not be related to the job position. These duties may not be written but implied in the instructions. They're duties that improve the quality of the job and help keep the team functioning at an optimum level. In most cases,

you have newer team members coming on board, don't assume that veteran team members will explain to the new recruits how meetings are supposed to be run or the best ways to ask for help; it's your job as a leader to set expectations and explain processes. If you don't make those norms clear for everyone, you risk creating an environment where people feel excluded, uncertain, or unwilling to contribute.

Set or clarify goals.

One of your most important tasks as a team leader is to set ambitious but achievable goals with your team's input. Make clear what the team is working toward and how you expect it to get there. By setting these goals early on, the group's decision making will be clearer and more efficient, and you'll lay the framework of holding team members accountable. Many managers inherit their teams, which often means they aren't creating new goals, but clarifying existing ones.

Keep your door open.

If there's one thing that new managers need to remember, it's that over-communicating in the early days is preferable to the alternative. "It's always better to start with more structure, more touch points, more check-ins at the beginning," How you do that—via big meetings, one-on-ones, email, or shared progress reports—will vary from team to team and manager to manager, but whatever the communication method, team members will never say "I wish the boss would communicate with us".

Score an "early win".

Identifying and solving problems that have a quick and dramatic impact early on shows that you can listen and get things done. Perhaps there is a longstanding employee frustration or an outdated work process. Maybe there is a project that you can easily fund or prioritize. Taking swift action demonstrates that "you are connecting and learning." But most importantly, achieving an "early win" builds team momentum. "It motivates people," "and can win you goodwill you might need later if the going gets tough."

Principles to Remember

Do:

· Be clear about what goes into your decision making and how you'll evaluate the team's progress

· Encourage team members to connect—better communication early on will help avoid misunderstandings and poor results later

· Look for roadblocks or grievances you can fix—it will earn you capital and inspire the team

Don't:

· Jump into trying to accomplish the work without building relationships with the team

· Assume that new team members understand how you or others work—take the time to explain processes and expectations.

· Be afraid to communicate often early on

Leadership Obstacles and Challenges

Common Leadership challenges in modern businesses:

1. Providing inspiration. As a leader, your team is looking to you to provide inspiration and motivation to complete their work. This can feel tough in a challenging work environment or if you're not feeling motivated yourself.

To inspire others, help your colleagues to focus on the value their work creates. Share the vision for the team and make sure each of them can connect to how their piece of work makes a difference.

Helping your team find purpose in their work is critical for employee engagement. In fact, 90% of employees said they would trade traditional reward mechanisms — such as extra pay — for greater meaning within their work

2. Developing others. A key part of your role as a leader is talent and employee development. It's important to search out the potential in your team members and encourage their growth. You'll need to understand their hopes for the future and find ways to help challenge and stretch them.

Make sure you take the time to listen to your team. Create formal and informal opportunities to talk about how they want to progress in their career and support them to take those steps.

3. Leading change. Change can often feel scary or uncertain and leading a team through it is a significant challenge for today's leaders. How we work is changing like never before and employees will look to leadership for guidance and reassurance. It's important to validate the feelings of your team and help them to feel their fears are heard. Helping them to find the positives of the change can lower resistance towards it.

4. Handling different perspectives. Workplace conflict can be extremely detrimental when handled poorly, causing stress to almost half (48%) of employees. There will be times as a leader when you must manage conflict between team members or between yourself and an employee.

Conflict can feel uncomfortable, but you need to solve it before it upsets the team.

If the conflict is between two employees, try and facilitate them in solving the issue themselves by encouraging listening and compromise. If one of your employees disagrees with you, make sure you consider their point of view and don't be afraid to change your approach if theirs is better.

Given the diversity of employees within the workplace, it's unsurprisingly that friction arises from individuals' differing experiences, ideas, and perspectives.

The challenge for leaders is creating space for those ideas to be shared and ensuring that conflicting ideas are channeled into a productive discussion that allows for growth and shared understanding.

5. Dealing with imposter syndrome. It's common, especially for new

leaders, to lack confidence or feel like they don't deserve to be in a leadership position. There are several different types of imposter syndrome.

You might question your own skills or judgment which could lead to indecisiveness. Or you might feel like you must tackle everything alone which could prevent you from asking for the help you need.

To squash imposter syndrome, look for evidence of your capability in feedback from your line manager or peers. You were given the role for a reason! And continue to develop your own skills in areas you find most difficult.

6. Managing a team. When you become a leader, you're either new to the team or you've been promoted from within it. Both things can be tricky. You'll have to build trust with new colleagues or manage a new dynamic with old ones.

In collaboration at work and with your new team, take the time to set expectations with each other. Agree on how you're going to work together, and how you prefer to communicate. If you've been promoted above your peers, don't just ignore that. Talk to them about how that feels and work through any frustrations they might have.

Communication challenges

1. Keeping everyone on the same page. In a fast-paced, changing environment it can be tough to keep everyone up to date with what's going on in the organization, or even in the team. Schedule quick, regular meetings to check everyone is receiving and understanding key information.

2. Staying honest. As a leader, you'll sometimes have to communicate difficult things to your team. It's important to share tough news or information as clearly and as honestly as possible. Prepare what you're going to say and make sure there's a chance for you to answer any questions.

3. Celebrating the wins. In the rush to make sure that work is being done, sometimes it can feel hard to make time to celebrate success. But it's key to keeping the team motivated and making sure their efforts are recognized.

Competency challenges

4. Making hard decisions. There will be times when you must make hard decisions. It might mean making an unpopular choice or taking an action that upsets someone but is best for the business overall. In those moments, you can feel alone and unprepared for the task ahead. Take a deep breath and remember that doing the right thing doesn't always mean doing the easy thing.

5. Managing resources. This can be especially tricky if you've got a great team that everyone wants to work with. Make sure you're clear on everyone's workload so that you know what requests for support you can accommodate and what you can't. Prioritize work coming in against the core purpose of your team so you make sure you say yes to the things that create the most value.

6. Delegating. In your eagerness to be successful in your new leadership role, you can end up micro-managing everything. It's important to learn to delegate, both to demonstrate trust in your team and to free up your time to concentrate on the most important things. If this feels hard, focus your attention on the most impactful tasks and ask for regular updates on those. Let the rest go.

7. Seeking feedback. It can feel strange to ask for 360-degree feedback from those you're leading but it's absolutely key to your success. Feedback is an opportunity to learn where you can continue to develop but is also a way to hear what your team does appreciate about your leadership. Set up your feedback mechanism to allow this and make sure to respond positively to whatever feedback you receive, it can take a lot of trust for employees to offer feedback in this way.

Internal challenges

8. Staying positive. Things won't always go to plan and staying positive in the face of disappointment is a real challenge for leaders. But it's important to move on, especially if the team is also feeling down and looking to you for motivation. Acknowledge the disappointment and identify what you can learn from it. Then use it as fuel to do better next time.

9. Wanting to be liked. There's no rule that says those of us with a tendency to people-please can't be leaders. But it does mean that sometimes things are going to feel a little uncomfortable. Some of the decisions you make as a leader may upset or frustrate people but that doesn't mean they're wrong. Do show empathy for the feelings of others but remember that, while being liked is nice, it is more important to gain respect for your leadership skills.

10. Staying calm. It's likely you'll experience times where you're under a lot of pressure in your role or you're having to lead through a crisis. During those moments it can be difficult to remain calm and focused on the task. Even if everything feels hectic, give yourself time and space to process information so that you can make decisions from a place of clarity, not chaos. Learning skills for emotional intelligence is important, but especially as a leader.

Tips to help you overcome any leadership challenges

1. Look after yourself. Leading others can be challenging. There may well be times when you feel daunted or overwhelmed by the responsibility. You'll need to be able to keep yourself motivated, deal with the stress that comes with uncertainty, and avoid emotional exhaustion and burnout.

Make sure you take the time to really switch off from work. Tempting though it is, don't read and reply to emails and messages outside work hours. It's exhausting for you and sets an expectation for your team to also work when they shouldn't.

At work, make sure you take a break in the day to eat and move. Self-care is important. Keep a bottle of water on your desk so you stay hydrated and try and take a screen-break every hour or so to give your eyes (and brain!) a chance to rest.

2. Get clear on your role. Make sure you know exactly what you and your team are responsible for, and what you're not! This really helps with managing resources and making sure you don't take on too much work and stretch yourself, and your team, too thinly.

Really understanding your role allows you to focus on the things that only you can do and delegating other tasks amongst the team. Delegating shows you trust your team to get things done and can help develop their skills.

And it frees up your time to concentrate on where you can add the most value.

3. Be authentic. There are lots of different types of leaders, and ways to lead, so don't feel you have to fit into a certain mold. Develop your leadership skills by trying different approaches to situations but don't try to be something you're not.

Being an authentic leader is important, now so more than ever.

Research says that employees value authenticity in leaders. It's ok to be human and look for support when things are challenging. In fact, it lets your team know that it's ok to be vulnerable and ask for help if they need it.

4. Find support. We all need help at times and, sometimes, leadership can feel lonely. Make sure you get the support you need so you can do the best job possible.

Leadership Tools

I do not publicly endorse any of the leadership tools listed below (one being better than the other), as I find them to be more of a guide, depending on your levels of experience and education. You must find what fits best for you.

Leader's Book: A leader's book is a physical (or digital) manifestation of the baseline responsibilities of being in charge. You must know your people, your equipment, and your mission. You must command the ability to describe each in detail at a moment's notice – and make decisions accordingly. The leader's book, when maintained appropriately, helps you accomplish that task.

Myers-Briggs Type Indicator: The Meyers & Briggs Foundation uses the theories of Carl Jung in the well-known Myers-Briggs Type Indicator (MBTI) test. The assessment determines your personality type from 16 different archetypes.

Harvard's Implicit Association Test: This test uncovers your unconscious biases – the attitudes and beliefs you have but may not be aware of. The assessment involves categorizing pictures and phrases and associating them as either "good" or "bad" as quickly as you can. Then after a brief survey, your results approximate your attitudes and assumptions.

Leadership Style Assessment: Most of us tend to have an idea about the kind of leader we are or would like to be. A leadership style assessment will tell you what kind of leader you are. You take a leadership inventory and work out your leadership style. Each style has its strengths but having the wrong perception of the kind of leader we are can be potentially very damaging, and this test will help to solve that problem by letting you know where you are now.

Leadership Blind Spot Assessment: A leadership blind spot assessment helps leaders identify areas of oversight across four major categories:

- Blind spots about yourself
- Blind spots about your team
- Blind spots about your company
- Blind spots about your markets

This is incredibly useful for most leaders to know where their existing blind spots are and where they need to improve.

Emotional Intelligence (EQ) Assessment: Emotional intelligence is now seen as the emotional equivalent of IQ since research concluded that emotional intelligence (EQ) counts for twice as much as intelligence quotient (IQ) and technical skills in determining individual success.

This test helps leaders to understand where they rank in EQ and to understand how it impacts their ability to manage, especially under pressure. An emotional intelligence assessment can help you determine where your emotional weaknesses exist when handling situations involving pressure, criticism, and failure.

Performing Under Pressure Assessment: Similar to the emotional intelligence assessment, the Performing Under Pressure self-assessment focuses on the circumstances that may cause leaders to underperforming when leading through a crisis.

The Performing Under Pressure assessment is geared towards the individual and how well you handle stress, anxiety, and criticism. Once completed, you get a score and brief description of strengths and weaknesses under pressure, as well as suggestions and where to improve.

LinkedIn Learning's Personal Effectiveness Tips: Formerly known as Lynda, LinkedIn Learning is LinkedIn's online training resource covering a multitude of topics and practical skills. One of the most impactful topics on there is personal effectiveness, and here LinkedIn Learning contains a great course on personal effectiveness by personal effectiveness guru Dorie Clark.

This course is practical, offered in bite-sized episodes, and includes some practical exercises to help your learning. As well as containing a section on identifying your own leadership blind spots, it also covers less common but very useful subtopics such as how to find a mentor and how to deal with change.

Clifton Strengths Assessment: This assessment is for managers and teams who are serious about improving their overall team effectiveness and dynamic. It consists of three steps:

1. Take the assessment – The test takes about an hour and contains 177 paired questions in which you must choose the one that best describes you.

2. Get your results – These will show your greatest talents and how to explain them.

3. Use your CliftonStrengths – Now you're more aware of your talents, you'll be able to identify opportunities that better match your abilities.

When you know the strengths and weaknesses of your team members, it becomes easier to assign tasks and challenge people in areas where they have room for professional and personal development.

Optimal Thinking Leadership Assessment: This tool is another leadership traits assessment. The assessment has been taken by an estimated 350,000 leaders from around the world including politicians, Fortune 500 executives, military leaders, and graduate students. The assessment consists of 25 statements in which you are required to answer "A" for "Always", "F" for "Frequently," "S" for "Sometimes," and "R" for "Rarely."

Once you take the test, you get some positive recommendations on areas where you could improve.

Define your expectations

Setting an example for the rest of the team or department: As a leader, your team looks to you for guidance on how to behave and perform their job responsibilities.

Holding everyone accountable, including yourself: Upholding leadership values and expectations shows your team that you hold yourself to certain standards, just as you have standards for their performance.

Establishing trust between leaders and employees: Leaders who manage their teams with respect and open communication may have an easier time building a trusting relationship with their team members.

Improving employee retention rates: Employees may be more likely to remain with a company or team if they feel supported and respected by their leaders. Managers and team lead who fulfill employees' leadership expectations might notice an increase in employee retention.

1. Be clear about your expectations

It's important to communicate clearly what you expect from your employees. Most employees appreciate when their managers or supervisors can tell them directly about their goals, needs and targets so that they know what to focus on and prioritize. While you may not always know the exact methods or actions needed to accomplish a specific objective, in these situations, you can still communicate to your team the overall scope or goal.

2. Set and achieve your goals

One expectation most employees have of their leaders is to model exemplary workplace behaviors, such as creating and actively achieving goals. Set and achieve goals, both for your own career and for your team.

Acknowledge and celebrate with the team when you or they reach those milestones. If you don't meet a particular goal, discuss honestly with your team what challenges you encountered and how you could optimize your processes for success in the future.

3. Offer respect to teams and individuals

It's important to show respect to your employees. Acknowledge when your team members put in significant effort and listen actively to their concerns or needs. If you demonstrate your respect for your employees, they're more likely to offer you respect in return.

4. Be humble

Be confident but humble about your career accomplishments. Most employees admire a leader who understands their own ability and capacity for professional growth. Show your team self-awareness of your weaknesses and your commitment to improving them. Acknowledging and taking steps to develop your weaknesses also models great workplace behavior for your employees.

5. Follow through with your commitments

Show good follow-through by adhering to your promises. Understand the steps a particular commitment involves before promising to do it, and then follow through with those steps. Showing your employees your ability to keep your promises helps to establish trust between you.

6. Motivate others

Inspire your employees to put forth their best effort, achieve their goals and develop their skills. Praise your employees when they perform well and help them find opportunities to learn new skills or advance in their careers. You might also offer your employees incentives to develop their careers or achieve team goals, such as catered lunches or additional time off.

7. Practice honest communication

Employees respect their supervisors more when they're open and honest about changes or recent developments within their company. Tell your employees this type of information directly. It's also important to encourage your employees to come to you with their questions or concerns.

There may be times when other company leaders don't authorize you to share certain information with your employees. However, in situations like this, it's best to be honest with your staff that you can't tell them rather than creating a lie or avoiding the truth.

8. Provide and be open to constructive feedback

Many employees appreciate the chance to learn from their supervisors or managers by receiving constructive feedback. Practice giving feedback that phrases critiques constructively and helps your employees achieve their full career potential.

Additionally, let your employees know you're open to their feedback as well. Show that you value their input, whether it's about current projects, strategies, or your leadership efforts. Establishing a relationship with your employees where you both feel comfortable giving and receiving feedback helps to build trust and respect.

9. Look for ways to improve

Continually look for ways to optimize your team's projects, methods, or policies. Effective leaders acknowledge their team's or their own successes while always striving to find or create ways to improve. Practice fostering a spirit of innovation in both you and your employees, such as by approaching familiar situations or procedures in unique ways.

10. Collaborate with teams

It's important to be an engaged team player with your employees. If you're a supervisor or other company leader, at times your role within a project or process naturally differs from your employees' responsibilities. However, the most effective leaders also attempt to be active members of their own team. You can still oversee your employees while participating in their projects and demonstrating skills such as collaboration, follow through and communication.

11. Help your employees advance their careers

As a leader, you can find and encourage your employees to pursue career development opportunities. Give your employees the resources, support, or encouragement to grow professionally. This might include budgeting for conventions or workshops relevant to your field, providing internal training on new technologies or giving them additional responsibilities that can help advance their skills.

Get involved, there are no hard-and-fast rules about when your involvement will have the most impact, there's no excuse for not taking responsibility when you see a problem growing. expect them to have the courage to deliver bad news. good managers generally step in under three types of circumstances: when somebody is falling behind in her commitments; when important personnel matters arise, particularly if there is conflict; and in a crisis.

The Price of Leadership

The true price of leadership is the willingness to place the needs of others above your own. Great leaders truly care about those they are privileged to lead and understand that the true cost of the leadership privilege comes at the expense of self-interest

· You will have to make hard decisions that will negatively affect people you care about

· You will be disliked despite your best attempts to the do the best for the most

· You will be misunderstood and won't always have the opportunity to defend yourself

Be prepared to become unpopular. People naturally resist change and view change in a negative way. Leaders take people through change. Initially, it might be hard, and your popularity will suffer, but your job as a leader is to make your followers see that the change is necessary and that it is all for the greater good. If you guide them properly and stick by them, your popularity will return once they see that you have their, and the organization's, best interest at heart.

Servant leadership is time-consuming. Servant leaders give up a lot of their personal time to invest in their subordinates' developing needs.

Your leadership journey can be a daunting process. Because you venture into the unknown with every team member and you are focused on empowering them and uplifting them, each person reacts differently which can become quite discouraging and overwhelming. It can become intimidating and scary sometimes because you know that it has become your responsibility for them to thrive

You will be forced out of your comfort zone. Since you are expected to work through difficult conflict situations and minimize the uneasiness of individual members, this type of leadership will take you out of your comfort zone. Conflict situations might become extremely challenging, but when you have developed a high EQ (Emotional Intelligence) you will increase the quick and positive resolution of the problem.

You must give a lot of yourself to your followers. Being truly authentic can become a bit invasive and it will be a real challenge not to lose bits and pieces of yourself throughout this journey. Looking after your mental wellbeing will have to become a strong priority when you are in a servant leadership position.

Servant Leaders concentrates on heightening collaboration skills, improving each of their employee's self-confidence, boosting their creative thinking, and increasing their risk-taking skills through developing proper decision-making abilities. Naturally, this will take huge effort and investment from the leader's side. But the rewards are amazing and at the end of the day, it will be worth it if you still look after yourself.

leadership can come at a cost and can be found in the lives of many people. We expect that our leaders don't scurry amid danger, but because of their position they put aside their own self-interest

A strong leader is someone who will sacrifice a little bit of time and a little bit of energy for the betterment of those whom they are serving

You deserve the team you get.

Once you become the leader, your days of complaining about anyone on your team are officially over. If you are unhappy, you can help your staff improve and grow. You can reassign them to align their strengths with a new challenge. You can even encourage them (strongly) to exit your organization. But you can't complain about them or their performance because you deserve the team you get. You are the person responsible.

You own it.

When you lead an organization, whether it is good, bad, or ugly, you must take responsibility for what happens on your watch. Blaming or pointing fingers is at best a sign of immaturity and at worst demonstrates a lack of integrity.

The fish really does stink from the head down.

Organizations are reflections of their leadership. So, when you see something, you don't like about your team, the first and best thing to do is pause in front of a mirror. Ask yourself, "What is it about me and the way I lead that is causing this behavior?" You will likely find that you are hiring people like you or that you are perpetuating patterns that exist in your life. Organizations are perfectly designed to get the results they are producing. If you want different results, you must change the design of the organization.

You can't outwit gravity

"Culture flows downhill." If you want to be at the top of the leadership mountain, make sure you are constantly looking out below to see what message you're sending.

Stretch The Goal

1) Knowing that others "impossible" doesn't need to be yours.

2) Train deliberately daily – what are you reading and/or doing daily to improve your leadership skills?

3) Understand that failure is inevitable, but it doesn't have to be the result; learn from it and adjust to try again – don't give up.

4) Great leaders know they can endure more than they think – don't hold back your abilities and don't let your mindset hold you back.

4 LEADERSHIP STRATEGY

Mission and Vision:

A Mission Statement defines the company's business, its objectives, and its approach to reach those objectives. A Vision Statement describes the desired future position of the company. Elements of Mission and Vision Statements are often combined to provide a statement of the company's purposes, goals, and values.

Clearly identify the corporate culture, values, strategy, and view of the future by interviewing employees, suppliers, and customers. Ensure that the objectives are measurable, the approach is actionable, and the vision is achievable. Develop buy-in and support throughout the organization.

Organizational Culture:

Organizational culture is the collection of values, expectations, and practices that guide and inform the actions of all team members. Think of it as the collection of traits that make your company what it is. A great culture exemplifies positive traits that lead to improved performance, while a dysfunctional company culture brings out qualities that can hinder even the most successful organizations.

Don't confuse culture with organizational goals or a mission statement, although both can help define it. Culture is created through consistent and authentic behaviors, not press releases or policy documents. You can watch company culture in action when you see how a CEO responds to a crisis, how a team adapts to new demands, or how a manager corrects an employee who makes a mistake.

<u>Self-Awareness:</u>

The single most powerful way to grow as a leader: Become truly self-aware.

This is, hands down, your biggest secret weapon in making yourself a great leader. I don't mean self-involved. There are lots of "leaders" who are way, way too focused on themselves, their evolution, their drama, etc. etc. Becoming truly self-aware means cultivating, daily, an accurate sense of how you show up in the world and what motivates you. For instance: What are your actual strengths and shot cummings as a leader and as a person? What impact do you have on others? What do you care most about? What's your moral compass, and do you use it as a guidance system? How closely do your actions line up with your promises? Without exception, the more self-aware someone is, the easier he or she is to coach; the more improvable and better able to accept what they need to improve.

Three ways to improve your self-awareness:

1) <u>Become a fair witness</u>. To be a fair witness means to report your experience as accurately and neutrally as possible. The more emotional attachment you must something, the more challenging it is to be a fair witness to that which you are involved. Reflect on your actions, your strengths and weaknesses, your mistakes, and successes, as though you are this impartial third party. What would he or she say about how you show up?

2) <u>Invite feedback</u>. People who want to be fully self-aware know that none of us can see ourselves entirely clearly without the aid of others. It's like trying to know what you look like without having a mirror. If you want to have a more accurate sense of how you are operating in the world, build a small group of people who know you well, see you clearly, want the best for you — and are willing to be critically honest with you in the service of that.

<u>3) Listen.</u> I believe strongly in the power of listening. I believe it is one of the foundations to success as a leader, as a parent, spouse, colleague, human being. And it's essential to true self-awareness. If you can learn to listen

fully, without filtering what you hear through your pre-existing notions, you will find that everyone around you is continually giving you clues – both subtle and overt – about how you're showing up, what they think of you, and how you're impacting them.

As you become more self-aware, you'll start to be able to improve both your strengths as a leader (so you can make best use of them) and those areas where you can improve….and move yourself up the bell curve to great.

Leadership Actions

1. Follow through and credibility matter: Be someone who delivers on their promises and leaves a positive impression on others.

2. Honesty and integrity: Always be honest and do what's right, even if it means admitting when we're not the right fit for a deal.

3. Lead from the front: As a leader, don't shy away from getting hands-on and doing the necessary work, leading by example.

4. Embrace humility and humor: Take on tasks outside your role, like manning the buyer's desk, and find joy in connecting with your team, even in light-hearted pranks.

These beliefs shape how I approach my professional and personal life, emphasizing the importance of integrity, leadership, and a willingness to roll up my sleeves and connect with others in meaningful ways.

Keep It Simple Leadership Principles: Taking your unique management style and organization into account, tools and services are provided to build a more productive, knowledgeable workforce. A workplace where employees are excited to go to work, managers seek their input and organizational goals are met. The guiding principles are:

Mutual Respect - Between Management and Employees.

Do, Don't Try - Making the decision to change is the easy part, the consistent follow through requires a daily renewal and commitment.

Hold Everyone to the Same Level of Accountability - When you don't hold everyone accountable for the behaviors you are seeking, then you get what you tolerate... When you accept rudeness, tardiness, and absenteeism you are saying it's OK. When you have different rules for different people you have poor morale.

Seek to Learn/Understand - Most people want to be heard. Learn to listen, not speak. Try to understand their point, and then decide if your point is still valid.
Mistakes Happen - Learn from them, embrace them, don't berate them.

Awards and Punishments: (Fair and Impartial)

Awards: Employee rewards and recognition have become a critical component of every successful business in today's date. It's unique way of adding value to your employees' achievements is what made it popular. And HR leaders are trying their best to leverage it further through new rewards and recognition ideas.

The concept of giving out accolades and rewards has been there for ages. When an individual is acknowledged for their good deed, they feel elated. This inspires them to keep getting better and binds them to greater goals.

Punishment: Nobody wants to be disciplined at work, but punishment for breaking rules ensures that the workplace is kept safe and productive. In fact, the US Occupational Safety and Health Review Commission states that an employer must make "a diligent effort to discourage, by discipline, if necessary, violations of safety rules by employees."

Exit Interviews: Exit interviews are important because they offer a deeper look at your workplace culture, day-to-day processes, management solutions, and employee morale. The purpose of an exit interview is to assess the overall employee experience within your organization and identify opportunities to improve retention and engagement.

Having a clear set of standards in place when conducting exit interviews can also play an essential role in risk management. When employee issues are addressed justly and effectively within the workplace, there is no need for external investigations, litigation, or bad publicity to follow.

Why are exit interviews Important

1. Departing employees are generally more forthcoming than those still in their jobs

2. You will learn the reason for an employee's departure (it may be different than you think!)

3. The exit interview allows the employee to provide constructive feedback and leave on a positive note

4. That last touchpoint provides you with an opportunity to review continuing obligations with the employee (e.g., non-competes, intellectual property agreements, etc.)

5. It provides the opportunity to ask if there are any open issues of which you need to be aware. This can help reduce risk and identify matters that may require immediate attention.

6. You will get a candid assessment of your organization's environment and culture

7. Insight into recruiting, on-boarding, and training needs may be revealed

8. The feedback will help you to identify areas that can help improve staff retention

9. Improvement opportunities in management development and succession planning can be detected

10. It's cost-effective to facilitate quality exit interviews

5 TOXIC LEADERSHIP

What is toxic leadership?

Toxic leadership is brought about by lack of self-awareness, lack of self-control and confidence, all of which are seeded by self-interest. As a result, a toxic leader's subordinates don't like him or her. Toxic leaders care all about themselves. They only think of their own feelings and disregard those of others. Somehow, toxic leadership is related to autocratic leadership style.

Such leaders act aggressively and treat others terribly. They distribute tasks in a restrictive manner and even try to intimidate their subordinates to preserve constricted control.

In my military career, as well as my civilian, we used a term to describe toxic leaders. We called them "seagulls", you know the bird that fly's in, shows up wanted, makes lots of noise, craps on everything, and then leaves. I am sure you have had a few "seagulls" in your careers.

Characteristics Of a Toxic Leader

<u>Self-destructive</u>

He only thinks that his own ideas are right. In every opportunity, he degrades his followers, and acts like a subordinate is of no use.

<u>Irritable</u>

A toxic leader is always irritable. He doesn't want to be bothered. He avoids and hates to be asked questions at any time. He doesn't allow an open and frank flow of ideas that would contribute to the betterment of the organization.

Arrogant/Narcissistic

Toxic leaders are arrogant. They are boastful and always think that they are right. They don't extend help to others. They hate to be corrected by other people, especially their team members.

Narcissistic leaders have an exaggerated sense of their own importance. They believe that they are special in some way. They have an excessive need for admiration from other people and lack empathy when dealing with others.

These leaders are self-serving. They are more interested in personal success and gain than in the long-term growth of the organization and the development of its people. They lack emotional intelligence and are disinterested in what their subordinates need. They will seek opportunities to build 'empires', taking on more responsibility if they feel it can make them look good.

Incompetent

Toxic leaders are incompetent, even though they always put in their mind that they are the best. They always criticize others.

Maladjusted

They are maladjusted and afraid of change. They are not flexible, and they are hard to deal with.

Lack of confidence

Toxic leaders are not confident within themselves. They also don't trust their subordinates. Every day we face problems. These problems must be solved right away to prevent them from pilling up. But with toxic leadership, subordinates will feel depressed and hopeless about the future

Tools of a toxic leadership

Organizational Structures

Toxic leaders can destroy organizational structures.

Workload

Workloads pile up. The reason for this is that toxic leaders are setting objectives that are unrealistic. In turn, this sets the company up to fail.

Corporate power structures

A toxic leader controls his or her team members.

These are the characteristics and tools of a toxic leadership. If you see these signs in your leader, then think twice if you're going to stay or leave. Toxic leadership degrades team members which lowers their self-awareness and self-confidence.

If you are a true leader, then you must assess yourself to determine if you exhibit any of these traits. If so, you need to internalize any issues you may have and take the necessary steps towards sincere change for the betterment of yourself as a person and as a leader.

Toxicity manifests in numerous ways:

Bullying and Commanding

Bullying involves things like unfair treatment, public humiliation, and other forms of threatening behavior. While some bullying is straightforward, other behaviors can be subtler yet still create toxicity. These include undermining one's position or responsibility, falsely taking credit, spreading rumors and half-truths, and social ostracism.

Whenever a leader commands, the power dynamic shifts and can become problematic. There is a thin line between commanding and bullying. Bullying can involve shouting, swearing, name-calling, malicious sarcasm, threats to safety, or actions that are threatening, intimidating, humiliating, hostile, offensive, or cruel. To cement their position, bullies evaluate performance unfairly, deny advancement, steal credit, attack reputations, give arbitrary instruction, and even assign unsafe work. They can interfere, sabotage, undermine, and encourage failure. The underlying phenomenon often identified as workplace bullying can result in physical as well as emotional and psychological disorders, including a diagnosis such as post-traumatic stress disorder.

Toxic Ambivalence: Sins of Omission

We can probably agree that toxic leadership does not necessarily require intentionality—it can be accomplished quite effectively as a sin of omission rather than commission. Simple ineptness and rank incompetence breed toxicity in their own way. Followers experience frustration where managers do nothing. Toxic Ambivalence refers to avoiding, delaying or deferring a decision until it has to be made, which really is a decision in itself. In some highly charged political situations the leader may walk a thin line between pragmatic indecision and toxic ambivalence.

Pretension: The Problem of Celebrity

Characterizing the leader as a celebrity may sound like an odd take, but we can see how aptly it fits. Through a variety of circumstances, any public event (physical, virtual or broadcast), we create celebrity. It is a form of leadership that emerges from visibility and branding. A name becomes increasingly recognizable, and a set of meanings is attached to it. There can then be a subtle shift, as a normal human being becomes a brand.

Uncritical followers, attracted by the brand, ignore a host of warning signs. A foundational indicator of toxic celebrity is a lack of accountability. Leaders enthralled with celebrity are literally "in thrall" to the unholy freedom to do exactly whatever feels good now. This is the dark side of celebrity. At the same time, celebrities are placed under extraordinary pressure without the benefit of being held responsible for results.

Hypocrisy: The Problem of the Servant Leader

Certainly, a servant-leader is sharply different from one who is a leader first, perhaps because of a need to assuage an unusual power drive or to acquire material possessions. The difference manifests itself in the care taken by the servant, first to make sure that other people's highest priority needs are being served.

Sometimes followers have a higher expectation for clear (or even forceful) direction than a leader does, which can confuse the leader in his or her attempts to be a servant. It can also cause a breakdown of communication and an abdication of responsibility for results where a leader attempts to be collaborative at all costs. Lack of direction, of management, of exercising leadership, can just as quickly create chaos among these types of followers and allow silos to fester.

Leaderless Leaders:

Leaderless leadership, these are the leaders who have a team of followers but have a great deal of difficulties exercising control and influence over them. There are some common reasons that make these leaders "leaderless". We usually blame the followers for making life difficult for the leaders, but it is the leaders who through their own doing, lose their personal powers and credibility in the eyes of their followers.

Key reasons that make a leaderless leader:

1. Lack integrity. Acting in ways that reflect no moral standing and do not know the difference between doing the right or wrong things.

2. Making poor decisions. It is not logic alone but also not making sound decision or hesitate to make them in a timely manner.

3. Poor planning. Disorganized in developing a definitive and clear work plan that provide clarity on what people must do.

4. Lack of guts to act. The faint hearted only invite disaster through their inaction and reluctance to move even though it is tough out there.

5. Lack management skills. Does not have a systematic and resourceful approach to get to the desired goal through well-aligned critical activities.

6. Unreliable. Not trustworthy, not dependable and require close supervision.

7. Lack loyalty to superiors and subordinates. Does not keep their people well informed, protect and defend their interests whenever required.

8. Lack enthusiasm and optimism. Has low energy, cheerless disposition,

and pessimistic outlook.

Leadership by email

Email presence, how you're perceived based on how you communicate through email, matters because it's an extension of your professional reputation and we know that your reputation is a key factor in your long-term career success. For better or worse, other professionals will judge you based on your emails. Your voice, your personality and your thinking are all on show. For that reason alone, email presence is important and should not be overlooked (even when you are in a hurry).

How to enhance your email presence

<u>1. Be aware that you are creating an email presence.</u> The first thing to do is be cognizant of the fact that every email you send creates an impression of you. This is especially true if the reader doesn't know you well already or hasn't yet had the opportunity to reconcile your in-person presence with your email presence.

<u>2. Re-read your email twice before sending.</u> I am as guilty as the next person, probably even more so, of not always checking my work before I hit send. I try to be conscientious with my emails but even then, I skip words, have misspellings, and could have punctuated better. Reduce these errors. If email is the only impression, you get to make – make it a good one. The next time you must send an important email – proofread, proofread, proofread, and then have someone else proofread it.

<u>3. If you are in rush, don't send it.</u> Save it (literally) for later. At times when you are rushed, it may be best to type out your thoughts and then save the draft. You can always send it later. Don't risk your email being taken in the wrong way, or being incoherent, because you just wanted to get it out. Plus, messaging apps have gotten much better for informal communication – use those if you are in a hurry.

<u>4. Check your exclamation marks.</u> A well-placed exclamation mark can make a big difference in the tone of an email. However, it's easy to go overboard. A nice rule of thumb is to use one exclamation mark per business email (if you use one at all). Too many exclamation marks can come across as juvenile, sarcastic, or even panicked!!!

5. Use an opening and a closing. A simple "hello" is a welcoming beginning and shows that you're thinking of the reader beyond the message you're out to deliver. A closing can serve the same function. Some people just rely on their electronic signature to close out their emails, others add a more personal touch by typing out a pleasant phrase followed by their name or initials. Be bold and change up your closing, an astute reader will realize that you take the effort to do so.

6. Know the "communication code". I like the analogy of thinking of communication as clothing. In your closet of communication options, email is business casual. If you were going to your office, you wouldn't dress up in your tux or evening gown, right? Hence, for email, you don't need all the formalities of an official letter. Similarly, you most likely wouldn't show up in your flip flops. Unless your office is at the beach, in which case we need to talk. So don't be too informal in email; don't use slang or overuse emoji's and cut down on the abbreviations you use when you text. Find a happy medium that's comfortable for you and has the right amount of formality. Find your business casual look.

7. Don't send an email. There are lots of times when an email might not be the best way to communicate. As mentioned, quick, informal messages are likely to be better received through a messaging app. Urgent information warrants a phone call first. Sometimes the best thing you can do for your email presence is not use it.

We've all worked with people that weaponize email and blast anyone who crosses it. But even level-headed leaders can self-sabotage with email if they're not careful.

Email was designed for ease of communication. But there are some things it was not designed for. Some situations need to be treated with phone calls or in-person communication, especially if they include difficult information or require more context.

There are at least three situations where leaders should almost never send an email:

1. To criticize. If your goal is to remediate poor performance or bad behavior, email is more likely to fail than succeed. Why? Because

it's easily misunderstood, it's one-dimensional, and it can even be dehumanizing.

Sometimes people need correction. It's just that email is a poor tool to produce the change we desire. Instead of improving the situation, we just might make it worse.

2. To complain. There's nothing wrong with complaining. Some situations are unacceptable, and it falls to us as leaders to point them out. Just not in email. The problems here are similar to those of the first situation. In both cases we might feel the need to document the error. That could be for HR or even legal purposes. But unless the case is hopeless or extreme, leaders should call first and follow up with an email to document later.

3. To deliver bad news. There was a time in my civilian career when we were facing being furloughed due to the government shutdown. We all knew it was coming, but we didn't expect to get the final notice in an email, and then a generic memorandum for record which we had to sign. Instead of addressing the staff in face-to-face meeting, they sent the bad news in an email.

   Email is excellent when it comes to informing, explaining, or even praising. But if it's not one of those, we should meet face-to-face if possible. If face-to-face is impossible, then we should set up a video conference or make a phone call. As an option, email should be dead last. Email seemingly allows leaders to avoid awkward conversations or difficult confrontations. But it's often an example of abdicating leadership.

6 COMMUNICATION

When it comes to business, clear, and effective communication from leadership offers a multitude of benefits to employers and employees. One such benefit is that it keeps employees aware of individual and organizational goals. When workers understand what is expected of them, they are more likely to be able to deliver the desired results, which can help increase job satisfaction, improve morale, and boost confidence.

Successful leaders must be able to inspire their employees. If leaders are not strong communicators, it will negatively affect the motivation and, consequently, the productivity of their teams. Conversely, when objectives and expectations are well-defined—and teams are collaborating to meet them—it can promote efficiency, enhance engagement, and improve overall work performance.

Strong leadership communication helps ensure that team members are aware of organizational challenges and opportunities, allowing them to make more informed decisions that ultimately benefit the bottom line.

Types of Communication

VERBAL COMMUNICATION

Verbal communication occurs when we engage in speaking with others. It can be face-to-face, over the telephone, via Skype or Zoom, etc. Some verbal engagements are informal, while others are more formal, such as a scheduled meeting.

Regardless of the type, it is not just about the words, it is also about the caliber and complexity of those words, how we string those words together to create an overarching message, as well as the intonation (pitch, tone, cadence, etc.) used while speaking. And when occurring face-to-face, while the words are important, they cannot be separated from non-verbal communication.

NON-VERBAL COMMUNICATION

What we do while we speak often says more than the actual words. Non-verbal communication includes facial expressions, posture, eye contact, hand movements, and touch.

NON-VERBAL COMMUNICATION (Cont.)

If you're engaged in a conversation with someone about an idea, it is important to pay attention to both their words and their non-verbal communication. They might agree with your idea verbally, but their nonverbal cues: avoiding eye contact, sighing, scrunched up face, etc. indicate something different.

WRITTEN COMMUNICATION

Whether it is an email, a memo, a report, a Facebook post, a Tweet, a contract, etc. all forms of written communication have the same goal to disseminate information in a clear and concise manner – though that objective is often not achieved. In fact, poor writing skills often lead to confusion and embarrassment, and even potential legal jeopardy. One important thing to remember about written communication, especially in the digital age, is the message lives on, perhaps in perpetuity. Thus, there are two things to remember: first, write well – poorly constructed sentences and careless errors make you look bad; and second, ensure the content of the message is something you want to promote or be associated with for the long haul.

LISTENING

The act of listening does not often make its way onto the list of types of communication. Active listening, however, is perhaps one of the most important types of communication because if we cannot listen to the person sitting across from us, we cannot effectively engage with them. Think about a negotiation – part of the process is to assess what the opposition wants and needs. Without listening, it is impossible to assess that, which makes it difficult to achieve a win/win outcome.

<u>VISUAL COMMUNICATION</u>

We are a visual society. Think about it, televisions are running 24/7, Facebook is visual with memes, videos, images, etc., Think about from a personal perspective – the images we post on social media are meant to convey meaning – to communicate a message.

We communicate continually throughout each day. We do it without thinking – we operate on communication autopilot. However, I encourage you to think about how you communicate. How do you communicate verbally? What nonverbal cues do you use when you are disinterested? Excited? Nervous? Are you a good listener? Can you write a concise, clearly articulated message? Are there barriers to how you communicate effectively?

Types of Communication Styles

Some situations will call for certain styles, or you might find that one style is particularly effective with one employee, while another works better for someone else. Styles can be combined, and people use styles different than their default one based on who they are communicating with.

<u>Assertive communication style</u>

This is considered to be the most effective communication style.

A person using this style is confident in their convictions but makes sure that they do not belittle or steamroll others in the conversation.

They do not resort to manipulation or pushing limits, rather they seek compromise and consensus through active listening and clearly expressing their wants or needs. Assertive communicators tend to have naturally high self-esteem, and they do not veer into passive or aggressive communication.

Calm, measured, and positive, many situations can be handled, and problems solved, by proactively using assertive communication.

How can you become a more assertive communicator?

· Speak from a place of confidence. If you're not a naturally confident person, fake it 'till you make it! You know best what you need to be successful, trust in yourself.

· Consider both your needs and the needs of those you are communicating with. Try to be solution-oriented, positive, and sensitive to the other person's point of view.

· Take a look at your verbs. Don't sugarcoat things or soften them. Say "will" instead of "could" or "should". Think of how different these two sentences sound: "I will take the lead on this project." vs. "I could take the lead on this project."

How can you effectively work with an assertive communicator?

· Give them the space to talk. They will let you know their ideas and their solutions; you just need to listen and give them the chance to communicate.

· Ask them what they think, and do not dismiss them if they are critical. If they think you are wasting their time, they will not bother being helpful to you.

· Expect them to ask for your thoughts as well. They want a conversation and are solution oriented. Bring your own solutions and criticisms and expect there to be a respectful conversation.

Aggressive communication style

This communication style can be hostile, threatening, and comes from a place of wanting to win at all costs.

An aggressive communicator behaves as if their contribution to the conversation is more important than anyone else, and the content of their message is often lost because of the tone of their delivery.

This type of communication can result in people feeling belittled, steamrolled, and intimidated.

In some cases, they may reactively push back at an aggressive communicator, not because the communicator is incorrect, but simply because the delivery of the message is so unpleasant that they instinctively disagree.

How to adjust your aggressive style to communicate better

· The confidence of an aggressive communicator is very close to the self-assurance of the assertive communicator. The big difference is that an assertive communicator values the other person's feelings, where an aggressive communicator does not. Switch your style by considering how your request - and the way that you deliver it - will affect the person you're speaking to.

· Take a look at your body language. Aggressive communicators use physical proximity, tone of voice, and gestures to underline their power in the conversation. This is not going to help your communication succeed. Step back, and tone it down. Colleagues will respond much better to a new, calmer conversational style.

· Change your point of view. You don't want to 'win' the conversation, you want to solve the problem. There's a much better chance of doing so if you work with others.

How can you effectively work with an aggressive communicator?

· This is a tough one. Aggressive communicators are notoriously difficult to work with. If you are unable to help them adjust their style, then prepare yourself for their blunt tone and unpleasant delivery.

· If you are in a position that you can advise the person on their behavior, introducing them to the communication styles, and the advice given above on adjusting it, could be a fruitful conversation.

· Get them down to business immediately, and try to streamline the conversation so that it doesn't wander off into unpleasant or irrelevant topics.

Passive communication style

This type of communication is also known as the submissive communication style. Another way of describing it is the "people-pleaser" type.

This type of communication is self-effacing, conflict-avoidant, and easy-going.

That's not to say that a passive communicator is always happy - in fact, this style of communication can lead to resentment building up over time because the person is unable to clearly communicate their opinions, needs, and wants.

They can find it difficult to effectively express themselves and want to avoid confrontation at all costs. This can directly lead to their good ideas never being heard, or for miscommunications to come up.

<u>How to adjust your passive style to communicate better</u>

· Be confident in the value of your contributions. If this is difficult for you, try the fake it 'till you make it advice we gave above for those hoping to become more assertive. Start small and build your confidence.

· Learn the value of "**No**". You can only take on so much work, and being able to set healthy workplace boundaries will help you be able to make sure your plate is not being overloaded by those looking to take advantage of your people-pleasing nature.

<u>How can you effectively work with a passive communicator?</u>

· Directly ask their opinions on subjects. Give them space within the conversation to offer their views and allow them the time to properly express them.

· Do not dismiss their ideas. They are working hard to build their confidence, if an idea won't work, point out the value of the idea and highlight the positive aspects of it, rather than cutting them down with a short "*Well, that won't work.*"

· Keep conversations positive and solution-oriented. Don't be angry or confrontational, that will cause them to shut down completely.

Passive-aggressive communication style

This style of communication, like the name suggests, combines aspects of both passive and aggressive communication styles. The passive exists on the surface, while the aggressive simmers beneath. Outwardly, the communicator seems sweet and easy-going, but they are operating from a place of anger and resentment. This bubbles up as sarcasm, being patronizing, starting rumors, or gossiping. This style tends to be very toxic in the workplace.

How to adjust your passive-aggressive style to communicate better

· Pinpoint where your anger comes from. Do you feel like your team doesn't listen to you? Do you feel like your contributions are overlooked?

· Look at your motivation when communicating. Are you coming from a place of wanting to help solve the problem, or are you lashing out? Adjust how you approach conversations and try to focus on positive, solution-oriented statements instead.

· Understand that business relationships can be complicated. Confidently and consistently approaching conversations in a positive and pleasant manner can improve your work life dramatically. You can't control how people act towards you, but you can control how you react to them. Changing the style of communication can bring a sea change to your relationships at work.

How can you effectively work with a passive-aggressive communicator?

· Try to understand their motivations, if possible. If there is an outside force causing them to communicate this way, then it could be within your power to help solve that problem. Do they consistently use this style when dealing with an aggressive communicator? Does this only happen when they are faced with short deadlines? Think about what is causing this to happen.

· Don't meet them with the same style. It can be tempting to respond in kind, but that continues an unhealthy cycle of communication. Instead, focus on assertive communication. If you offer responses that highlight their contribution, offer win-win solutions, and do not belittle them, they will see the value of a different type of communication style.

Focus on their message and rephrase it in an assertive manner. Change the tone while demonstrating the communication style you would prefer their message to be delivered in.

Manipulative communication style

This style of communication uses cunning, deceit and influence to control the outcome of the conversation, and the actions of the people around them.

Manipulative communicators rarely say what they mean, instead they will bury their real goals within layers of obfuscation to get their way without the other person even realizing it. This style will lead to some clashes within teams or departments if one employee is using it consistently.

How to adjust your manipulative style to communicate better

Work on being more direct with your wants and needs. Practice asking directly for what you would like, and having grace if the answer is not what you would like.

Understand that this style of communication quickly breeds resentment among colleagues. Be consistent in your change, and remember to look for solutions that benefit everyone, not just you. Stay away from using emotional arguments. Be factual, precise, and direct in your communication.

How can you effectively work with a manipulative communicator?

As with an aggressive communicator, you should endeavor to steer them into assertive communication. You could try redirecting them when they try to use emotional arguments, not allowing them to derail the conversation towards their own goals and doggedly sticking to the topic of the conversation.

Don't respond to them with criticism, rather use the assertive restating tactic to clarify what they are saying while demonstrating the way you wish them to say it. You will have to be vigilant with these types of communicators that they do not sneakily introduce their agenda into conversations.

Use your patience and stay calm during conversations. Stand strong in your convictions and they will see that you are not an effective target for their communication style. By firmly but politely refusing to allow them to manipulate you, they will adjust their approach.

Communication challenges

1. Keeping everyone on the same page. In a fast-paced, changing environment it can be tough to keep everyone up to date with what's going on in the organization, or even in the team. Schedule quick, regular meetings to check everyone is receiving and understanding key information.

2. Staying honest. As a leader, you'll sometimes have to communicate difficult things to your team. It's important to share tough news or information as clearly and as honestly as possible. Prepare what you're going to say and make sure there's a chance for you to answer any questions.

3. Celebrating the wins. In the rush to make sure that work is being done, sometimes it can feel hard to make time to celebrate success. But it's key to keeping the team motivated and making sure their efforts are recognized.

Feedback:

Oral: effective tool as it can be provided easily in the 'teachable moment' and in a timely way.

Written (informal, formal): provides staff with a record of what they are doing well, what needs improvement and suggested next steps. It needs to be timely, written in a manner that is understandable to the staff and actionable so that the staff can make revisions.

Verbal and Non-verbal: outbursts, facial expressions, etc.

<u>Peer and self-assessed feedback</u>: Peer feedback occurs when staff offer each other advice and suggestions in relation to each other's work. Self-feedback must be taught explicitly to ensure staff have the skills to apply this to their own work.

It is the quality of feedback that counts.

7 PROFESSIONAL DEVELOPMENT

Professional development unlocks value at all levels. For the individual, this means honing their skills, learning new tools, and growing as a professional, whether that means focusing on leadership or broadening and deepening capabilities in their career. As individuals develop, organizations flourish. Engagement, alignment, and effectiveness are unlocked. At the leadership level, professional development trains leaders to embrace complexity and ambiguity, to tackle the unsolvable.

Two basic planning steps in preparing for staff development include:
1. Reviewing policies, regulations, and procedures
2. Conducting a needs analysis

Three recommendations are:

1. Let team members lead - empower staff to accomplish their jobs more effectively and to grow as professionals, put staff in charge of training functions and/or planning. (Not everyone will want too, many are content just sitting in the back, not willing to participate or contribute) So how do you develop these people?

2. Make staff development serve wider goals (it doesn't necessarily have to be within the section/group, (don't worry about losing good people to promotion, remember you are helping to develop their career, not just yours)

3. Organization statements of vision, mission and planning need to be reviewed - and possibly revised – annually. (You don't need to recreate the wheel, but you can update the spinner from time to time).

Characteristics of Successful Staff Development

· Planned by staff groups that include representatives of all stakeholders.

· Task-specific, focusing on department or administrative goals - including those identified by a staff needs assessment.

· Led by staff members who are proficient in teaching adults and in modeling effective teaching strategies.

· Adequate in the time it allows for staff to learn, practice, and apply new concepts and techniques.

· Sensitive to staff members' personal needs and schedules.

· Flexible in allowing staff to use what they learn in a variety of ways appropriate to their individual situations.

· Supported by handouts that spare trainees from having to take copious notes and that trainees can use to support lessons.

· Appropriate to staff members' current attitudes and expertise.

· Supported at the Organizational and Department levels with adequate, ongoing staff and time for person-to-person and small-group instruction.

· Accepted throughout the organization as an ongoing activity, not a single event.

· Continuously evaluated and improved.

Staff Development Needs Assessment

People

· What knowledge, skills, and attitudes do current staff members have?

· Has the organization surveyed staff to learn what staff think?

· Will there be any incentives that might be effective?

· How many "experts" are available to present staff development programs and act as coaches?

Money

· Is there a need for funds, or current budget and sources for staff development?

Equipment and Supplies

· What equipment and supplies are needed or already on hand?

· Where is equipment for staff development located (or where might it be)?

· What changes in facilities, furniture and furnishings, and hardware accessories will be needed to ensure staff health, safety, and comfort while using new technologies?

8 PROFESSIONAL DEVELOPMENT PLANNING

A professional development plan documents the goals, required skill and competency development, and objectives a staff member will need to accomplish to support continuous improvement and career development. A professional development plan is created by the leader working closely with the staff members to identify the necessary skills and resources to support the staff member's career goals and the organization's business needs.

Professional development for staff members begins when a new member joins your team. In addition, all staff members should have a "living" professional development plan in place. Planning should not take place only after a staff member is identified as needing improvement. Professional development plans should be reviewed on an on-going basis throughout the year, with at least one interim review discussion between the staff members and supervisor prior to the end of the yearly performance review period.

Professional Development Planning Steps

Use the following steps to create a professional development plan with your staff.

- Step One: Request a self-assessment from the staff member
- Step Two: Develop your assessment of the individual's skill level

- Step Three: Assess the department and organization's needs
- Step Four: Explore development opportunities with the staff member
- Step Five: Record and analyze the staff member's progress

<u>Step One: Request a self-assessment from the staff member</u>

Have the staff member complete a self-assessment of their interests, skills, values, and personality. When evaluating the staff member's responses, keep these questions in mind:

- What skills, career opportunities, technologies interest the individual?
- Do those skills/interests/goals support the organization's needs and goals?
- What are the short- and long-term steps to get there?

<u>Step Two: Develop your assessment of the individual's skill level</u>

Based on the staff member's self-assessment, their work record, and your own observations, determine the staff member's skill level in the following categories:

- Technical skills: skills needed to get the job done.
- Social skills: how do they work with others?
- Aptitudes: natural talents; special abilities for doing, or learning to do, certain kinds of things.
- Attitude: outlook, feelings, mind-set, way of thinking, and point of view.

<u>Step Three: Assess the department and organization's needs</u>

For professional development to be successful, the staff member's needs and interests must be applied to address organizational objectives. The staff member's career path must align with the organization's workforce needs. In creating a professional development plan, consider the following goals:

• Departmental goals

• Team goals

• Individual goals

Step Four: Explore development opportunities with the staff member

Professional Development Academy - The Professional Development Academy is a center dedicated to providing professional development training programs and resources for staff that support identified staffing needs. The Academy offers long-term training programs with a tailored curriculum designed to develop skills and capabilities needed to fill positions within the organization.

New Projects & Responsibilities - Explore what new projects and responsibilities the staff members can assist with in their own department. Staff members can use such opportunities to develop new skills such as web design, business writing, and project management.

Workshops & Seminars - Learning & Organizational Development offers a variety of workshops and seminars that help a staff member develop their work and computer technology skills.

Educational Opportunities - There are a variety of educational opportunities available throughout professional organizations, community projects, and universities.

Volunteer Opportunities - Volunteer opportunities can present a unique way for a staff member to develop certain professional skills.

Mentorship - Interested staff can be paired with mentors for a variety of activities including information interviews, shadowing, tutorials, etc.

Step Five: Record and analyze the staff member's progress

Collect feedback from the staff member about their development progress to assist in identifying what the staff member is doing well, build on their skills, correct any problems that may arise, and help them develop new abilities that will improve personal performance as well as organizational outcomes.

Use a Performance Log for tracking, recording, and providing feedback from the staff members. Record dates, events, expectations, and the impact of action steps on their development. Make sure to record:

• Observations of enhanced skills or knowledge and how they were applied.

• Progress towards goals and objectives.

• Observations where skills / knowledge could be applied - used for future discussion.

Professional Development Suggestions

Building a Professional Development Plan

A professional development plan is created by the leader working closely with the staff member to document the goals, required skill and competency development, and objectives the staff member will need to accomplish to support continuous improvement and career development.

Expand Accountabilities:

• Expand a staff member's job content to include a wider variety of tasks, risk taking opportunities, greater responsibility and/or authority.

• Ensure that different types of tasks and levels of responsibility are needed, not just more work in the same task.

• Monitor and review the staff member's progress.

Attendance and involvement in Department, Division, Staff or other specially designed meetings:

• Select the meeting(s) which will provide the staff member with the knowledge, exposure or broadening needed.

• Pre-plan the staff member's involvement in the meeting and determine the expected outcome for the staff member.

• Review the meeting with the staff member - emphasizing the areas related to the anticipated results (e.g., exposure to thinking at a more global instead of local level).

On the job coaching and feedback counseling:

• Provide skill and knowledge by answering questions, reviewing work, discussing progress, and providing general direction.

• Work through actual problems to share insight and problem-solving strategies.

• Monitor development progress and feedback on the results.

Serving as a conference leaders or trainer:

• Schedule the staff member to serve as a conference leader or trainer.

• Teaching others is a valuable way to obtain new knowledge, broaden one's viewpoint and clarify one's thinking.

• Select training responsibilities that put the staff member in touch with new operations, process, or operating problems.

• Review audience responses with the staff member.

Formal presentation Opportunities:

• Assign a presentation that will give the staff member an opportunity to develop planning and organization skills as well as effectiveness in speaking.

• Review and critique the plan prior to presentation.

• Provide feedback on the effectiveness of presentation.

Understudy Training:

• Prepare the staff member to assume the responsibility of his or her supervisor either as a substitute or as a replacement.

• Provide opportunity to observe and be involved in the day-to-day responsibility of the supervisor's job.

Planned exposure to another function:

• Expose the staff member to lateral job functions either on a project or observation basis.

• Identify and plan the expected outcomes with the staff member.

• Review and critique the exposure experience.

<u>Lateral Transfer</u>:

• Transfer the staff member to another job at the same position level to broaden his/her knowledge and experience as well as develop an appreciation for viewpoints and interrelationships of various groups within the organization.

<u>Vacation or replacement assignments</u>:

• Assign the staff member to responsibilities of his/her superior when the superior is on vacation or extended travel.

• Review the assignment with the staff member and critique his or her performance.

<u>Special assignment</u>:

• Select a problem area that will require increased analytical ability, improved judgment, and knowledge of the organization and/or decision-making skills.

• Assign an in-depth study of the problem area.

• Review and critique the study.

<u>Special team projects or assignments</u>:

• Place the staff member in a group problem-solving situation to develop his/her skills in working with others.

• Place the staff member in a group with representatives from various parts of the organization to develop the participant's knowledge and understanding of the organization.

• Review and critique the staff member's performance or participation.

Work one-to-one with a specialist:

• Arrange to have the staff member talk or work with people experienced in specific areas.

• Arrange to have the staff member talk or work with external professionals.

• Follow-up to review new ideas, perspectives, and knowledge gained.

In-house training programs:

• Enroll the staff member in an applicable in-house course or program at Learning and Organization Development, the Professional Development Academy and/or Continuing Education.

• Have the staff member report on what was learned.

• Provide the staff members with access to computer-based training.

Outside functional training:

• Carefully select outside professional development programs.

• Ensure that the programs (e.g., leadership development, functional broadening, or technical education) meet the organization's and staff member's specific needs.

• Consider community college and university courses that can be used to teach specific skills.

Self-development Programs:

• Schedule participants in interpersonal skills and motivational programs.

• Expect only short-term effects.

Study manuals, Bulletins, Reports, and other printed materials:

• Carefully select the reading material and have the staff member study it.

• Follow-up and discuss the insights and/or knowledge gained as well as its applicability.

Mentoring:

• Share your experience, knowledge, and ideas on specific topics with the staff member to encourage him/her to try new approaches to solving problems.

Home study courses and formal evening courses:

• Select courses through many community colleges, universities, and professional associations.

• Such courses require a great deal of self-discipline and effort.

• Review progress.

Active participation in community and civic affairs:

• Encourage the staff members to become involved in community organizations to develop leadership ability, skills in planning and organizing, problem-solving ability and community awareness.

Personal development workshops:

• Carefully select and evaluate the workshops.

• Select participation on a voluntary basis only.

• Select workshops to help gain new insights into his/her personal style and how others perceive him/her in interpersonal relationships.

Modeling or Simulation:

• Use of simulation to define skill levels, improve individual and team skills

9 TEACH, COACH, MENTOR, COUNSEL

We often throw the phrase "teach, coach, mentor" around, but don't take the time to unpack these leadership responsibilities and think about how we'll fulfill the role of teacher, coach, and mentor. XXX

Teach, Coach, Mentor, Counsel

Teaching

Teaching is perhaps the most well-known way to help others to learn. Teachers also need to be skilled motivators, using both short-term and long-term motivational techniques.

Skills Needed for Teaching

As well as subject knowledge teachers should:

1. Enjoy communicating your understanding to others.
There is an element to most teaching.

2. Have confidence.
You will need the confidence to look calm and professional even when tired and stressed.

3. Have great organizational skills.
Have you prepared for the training and done rehearsals?
Have you kept what is needed for anyone in the group who was away?
If you are part of a teaching organization, have you fed back results to any interested colleagues?

4. Work effectively in groups.
Define what is to be taught and how to deal with any difficulties.

5. Be able to deal with conflict.
There may be staff who need to be told to work harder, or a disagreement between staff that you need to help to sort out.

Types of Conflict

There are three types of conflict, personal or relational conflicts, instrumental conflicts, and conflicts of interest:

Personal or relational conflicts are usually about identity or self-image, or important aspects of a relationship such as loyalty, breach of confidence, perceived betrayal, or lack of respect.

Instrumental conflicts are about goals, structures, procedures and means something tangible and structural within the organization or for an individual.

Conflicts of interest concern the ways in which the means of achieving goals are distributed, such as time, money, space and staff. They may also be about factors related to these, such as relative importance, or knowledge and expertise.

6 . Motivate your staff to do their best.
This may require encouragement and/ or criticism, and probably a bit of both at different times.

7. Empathize with your Staff.
If you can see that your staff are exhausted, there may be no point in trying to teach a very complicated topic. You need to create a feeling that you are all working together towards the same goal. This means building up trust and rapport.

8. Give feedback.
Whether this takes the form of comments on performance or marking written work, it needs to be constructive. Offer praise as well as criticism whenever possible and tell your staff how they can improve.

The Best and the Worst of Teaching

Best that can Happen
•You get to tell enthusiastic people about a subject that you love.
•You may have interesting discussions that push you to think on your feet and expand your own understanding.
•You see less-able staff blossom and manage to get the outcome they need.
Worst that can Happen
•Lots of outside work may be needed to evaluate work or prepare sessions.
•Not all your staff will be enthusiastic. People can be extremely rude, and many adults that you teach may have their own views on things.

Coaching

The coach does not necessarily have to be an expert in the subject matter, but someone who specializes in helping to unlock the potential of others. The idea is that individuals hold the key to their own success, and do not need others to tell them what to do.

Coaching is a process that aims to improve performance and focuses on the 'here and now' rather than on the distant past or future.

While there are many different models of coaching, here we are not considering the 'coach as expert' but, instead, the coach as a facilitator of learning.

There is a huge difference between teaching someone and helping them to learn. In coaching, fundamentally, the coach is helping the individual to improve their own performance: in other words, helping them to learn.

The essential part of coaching is to help people to learn to silence that inner voice and allow their instincts, or their subconscious, to take over. Sometimes that means distracting it, and sometimes it's about exploring the 'worst case scenario' and removing the fear.

The Competence Cycle Model of Learning

1) Unconscious Incompetence

You don't know that you don't know about something.

2) Conscious Incompetence

You have become aware that you lack a particular skill.

3) Conscious Competence

You have learned how to do something, but you still need to think about it to do it.

4) Unconscious Competence

You have learned how to do something so well that it has become hard-wired into your brain.

Coaches need to identify the stage at which an individual is at, to use the right sort of language to help them move to the next stage. After all, it's difficult to try to improve a skill if you don't know that you lack it.

Coaching and mentoring require some very specific skills, particularly focused on facilitating and enabling others, and building good relationships.

Mentoring

Key Aspects of Mentoring

Unlike a management relationship, mentoring relationships tend to be voluntary on both sides, although it is considered possible for a line manager to also be a mentor to the people that they manage. Unlike a coaching relationship, mentoring relationships are more usually unpaid.

The idea behind mentoring relationships is a semi-charitable one: that the more successful, senior partner, the mentor, wishes to pass on some of what they've learned to someone else who will benefit from their experience.

Some organizations run formal mentoring programs that match mentors with learners. However, less formal mentoring relationships can also work well.

Some mentoring programs tend to have four key elements: improving performance, career development, counselling and sharing knowledge. There is also an element of the mentor acting as a sponsor for the learner.

Mentoring relationships, especially formal ones organized through a mentoring program, are often entered into with a defined time limit, or a defined goal. Having such a framework in place can be easier for both parties to agree on than an open-ended commitment.

Types of Mentors:

Adviser or Information Resource

This is a role that is often used when someone is new to an organization, during an induction period for example. The mentor helps the learner to develop their understanding quickly, or to support their ongoing career development. In this case, the mentor becomes a valuable source of information, and not just a sounding board. The mentor may also share his or her experience to help the learner to understand a particular work situation.

Role Model

This role, interestingly, is the one that requires least effort from the mentor, because it is usually about how they behave naturally.

The learner may have been attracted to them as a mentor because of the way that they handle certain situations. The learner will therefore learn from watching how the mentor behaves, both in the mentoring relationship and beyond.

Critical Friend

The role of critical friend is one of the most important, though most difficult, mentoring roles to successfully undertake.

It requires the mentor to listen, encourage, draw out, reflect back and challenge assumptions, and, if necessary, provide critical feedback on ideas or plans under discussion.

The role requires giving constructive feedback, and strong emotional intelligence and awareness of feelings.

The Benefits of Mentoring

A mentoring relationship can have huge benefits for both parties.
For the learner, there is obviously the opportunity to explore their learning and benefit from someone else's focus and expertise, either in a particular subject or in supporting the learning process.
Learning and development can often get pushed to the bottom of the 'to do' list when we're busy, and a mentoring relationship brings it forward again, not least because of the need to prepare for and then attend a mentoring session.
For the mentor, the benefits may be more subtle.
It is always nice to feel that we are doing something valuable and supporting someone else. A mentoring relationship may also be a useful opportunity to work on a leadership style, particularly coaching, or other communication skills without comments from co-workers about the change.

Counseling

Counselling is perhaps more intense than coaching or mentoring, and often considered therapeutic rather than supportive. But counselling nonetheless shares some characteristics with the other ways to help others learn the position that the learner holds the answer to their own problem, and the desire to help them take responsibility for that.

Counselling is:
The process occurs when a staff member and leader set aside time to explore responsibilities and opportunities to learn and develop their careers.
The act of helping the staff to see things more clearly, possibly from a different viewpoint. This can enable the staff members to focus on learning experiences and behavior, with a goal of facilitating positive change.
A relationship of trust. Confidentiality is paramount to successful counselling.

Counselling is not:

- Giving advice or being judgmental.
- Attempting to sort out the problems of the staff member.
- Expecting or encouraging a staff member to behave as the leader would if confronted with a similar problem in their own life.
- Getting emotionally involved with the staff member.

- Looking at a staff member's problems from your own perspective, based on your own value system.

The Role of the Counsellor

First and foremost, counsellors need to be aware that no two people are alike. No two people understand the same language in the same way; their understanding will always be linked to their personal experience of the world. The role of the counsellor, therefore, is to help the staff members develop their own understanding of their situation.

Counselling Skills

There are several skills that are required by counsellors. Perhaps the most important are good communication skills.

Counsellors need to be particularly able to listen effectively, giving their full attention to the staff. They need to be aware of body language and other non-verbal communication. Staff will often communicate far more non-verbally than verbally, so this is an important area of skill.

Questioning is an important skill for counsellors, just as it is in coaching. Counsellors use questioning both to improve their understanding (as a form of clarification), and as an active way to help expose the staff's feelings and emotions. They will also use reflection to show that they have heard the staff member and to validate their feelings and words.

Counsellors also need to be able to build a certain amount of rapport with their staff members, but not to an extent that would allow them to become emotionally involved.

Differences between Teaching, Coaching, Mentoring and Counselling

Although teaching, coaching, mentoring, and counselling all share some key characteristics and skills, they are nonetheless quite different and it's important to be aware of the differences.

Teaching and Training

Teaching and training involve an expert teacher who imparts knowledge to their staff.

Although the best teachers will use participative and interactive techniques, like coaching, there is very definitely an imbalance of knowledge, with the teacher as expert knowing the 'right answer'.

Coaching

Coaching involves the belief that the individual has the answers to their own problems within them.

The coach is not a subject expert, but rather is focused on helping the individual to unlock their own potential. The focus is very much on the individual and what is inside their head. A coach is not necessarily a designated individual: anyone can take a coaching approach with others, whether peers, subordinates, or superiors.

The key skill of coaching is asking the right questions to help the individual work through their own issues.

<u>Mentoring</u>

Mentoring is like coaching. There is general agreement that a mentor is a guide who helps someone to learn or develop faster than they might do alone.

In the workplace mentors are often formally designated as such by mutual agreement, and outside of an individual's line management chain. They usually have considerable experience and expertise in the individual's line of business.

A mentoring relationship usually focuses on the future, career development, and broadening an individual's horizons, unlike coaching which tends to focus more on the here and now and solving immediate problems or issues.

<u>Counselling</u>

Counselling shares some characteristics with the other ways to help others learn the position that the learner holds as the answer to their own problem, and the desire to help them take responsibility for that.

Counseling is a process designed to inform, develop, and hold accountable, a staff member's actions and responsibilities. It involves a relationship of trust and is more than annotating deficiencies and negative feedback.

10 POETRY OF LEADERSHIP

It may come as a surprise to you, if you read my bio, and see I am a retired Infantry Soldier. I served over twenty-four years, with multiple training deployments and six combat tours in Iraq and Afghanistan.

After retirement, I still work and teach in a government facility as a civilian. I have been writing poetry for many years and have published ten books so far.

I use my poetry as a way to paint a picture, that most books and manuals cannot produce. I hope you will find some insight and learn something new in each one.

Character to Destiny 7 SEP 23

It is not a matter of chance,
But of choice,
You must be strong enough to face the world,
Or be crushed by its fears,

Other's opinions are not your purpose,
And your history is not your horizon,
You must be a rebel with a cause,
And an angel in temperance,

The winds of change blow where they blow,
Sometimes with gust and fury,
And others on a gentle breeze,
But destiny does not make house calls,
You must go out and find it,

Your imperative is to follow it,
Accept it,
No matter where it leads,
For there are no secret doors,
That lead to your purpose,

Impose your own plan of action,
Confront darkness with the illumination of truth,
For your reputation is what others think of you,
And your character is what God knows of you,

Never forget,
Whether it be storm or fire, peace, or grace,
The sword of destiny has two edges,
And you are one of them,

Dream outside the lines,
And find your path,
The one unknown to many,
For your character,
Will lead you to your destiny.

Approachable 30 May 24

Approachability, opens the doors for opportunity,
It empowers others to take ownership,
Build rapport,
And project competence in skills yet discovered,

Approachable leaders are accessible,
There doors are always open,
They are consistent in their display of emotions,
And create predictable and consistent outcomes,

Approachable leaders,
Are emotionally intelligent,
They display an attitude of acceptance,
Encourage people to talk,
And engage them when they do,

Unapproachable leaders,
Create environments that prevent people from bringing their best,
Dismissing their strengths and talents,
Challenges and solutions,

Unapproachable leaders,
Withhold acknowledgement and response,
They are often distant and preoccupied,
Apathetic, and disconnected from those they serve,

Some are brilliant,
Their minds are closed,
But their mouths always open,
They cannot, and will not see all views,
Or make sense of simplicity,

You will never learn the truth about your leadership,
If everyone is harmonizing everything you do,
Be willing to hear, what you don't want to hear,
And all that you need to hear,

Remember, the image you present,

Is the image of your reality to others,
Take your ego out of your leadership,
Or it will lead you astray.

Self-Awareness 30 May 24

When you look in the mirror,
Into your own eyes,
Do you see the real you,
Do you see what you wish to see,
Or a glimpse of something you wish wasn't there,

It takes courage,
To endure the pains of self-discovery,
To develop your own conscious,
And become responsible for first creations,

Everything you are looking for,
Is not out there, but within,
Confront the doubt,
And banish it with forgiveness,

To know yourself,
Look at your fears,
Fear itself is not important,
But it points you in the direction,
Of things that are worth breaking boundaries,

Never try to do anything,
That is outside of who you are,
Look in the mirror and see,
The truth of who you are,
And that which you hide within,

Listen to the whispers of your soul,
Find the vocation of yourself,
For you cannot escape reality,
Or the noise of others in your head,

Every leader has a true vocation,
To find yourself, discover your destiny,
To live resolutely within,
Anything else would be existence,
Acceptance of the aggregation's ideals,

And conformity of fear.

Vision 24 May 24

Leaders create their vision,
Articulate their vision,
Passionately own their vision,
And relentlessly drive it into completion,

A leader's vision comes from courage,
Not position,
From the example they bring,
Not their authority,

A leader's vision looks forward,
And sees what others do not,
What the organization could be,
Achieving now, for the future,

Leaders go to places,
Where they want you to follow,
And wherever they take you,
There you will be,

Good leaders,
Create great leaders,
Leaders must create a vision unborrowed,
That can translate into reality,

A leader's vision,
Must speak the language of their staff,
For it is a call to become something more,
A promise,
That inspires energy that can change the world

A leader's vision will always meet opposition,
From those with the weakest minds,
Leaders must continually learn, and develop themselves,
Before operating beyond their own borders,

The greatest danger for any leader,
Is not that they aimed to high, and missed,
But that that they aimed too low,
And reached it.

Honor 28 May 24

We value virtue,
But do not discuss it,
For our attention is turned to the ratchets,
The malefactors,
And the fear of abrogate perceptions,

An honorable leader,
Is one who is truthful,
Free from deceit,
Above cheating, lying, stealing,
One who will not bend their politics,
To their purses,

A leader's character,
Is judged on how they keep their word,
The example they set,
Not on the backs of blood and bone,
Nor the monarchy they try to build,

If you are the one to pass sentence,
You must look into their eyes,
And hear their words,
Do not defer judgement,
Or the responsibility of your position,

Do not abandon your traditions,
You must stand up to the darkness,
For true honor comes from the heart,
Not the opinions of others,

Do not sacrifice your honor,
For the sake of victory,
For only moral victories,
Are true victories,

Good leaders, do not become legends,
They do not seek to become legend,
For they know their actions,
Will outlast their mortality,

An honorable life,

Is a complete life,
a life with no excuses or compromises,
But if you choose to concede your honor,
It is almost impossible to regain it.

Humility 28 May 24

The first test of a leader,
Is their humility,
For they know greatness is not of them,
But through them,
They see something divine in others,
And are endlessly merciful,

A leader is at their best,
When the mission is completed,
The vision is seen,
But the credit goes to "We",
And not "Me",

Leaders must learn,
What questions are unanswerable,
And not to answer them,
And what to answer,
When truth will impact reality,

There is nothing more deceitful,
Then the appearance of humility,
For it is often only carelessness 0f opinion,
And hinted displays,

Let no experience be wasted,
Discover merits where you least expect them,
And talents in those,
You thought were impossible,

Nurture your mind,
Beyond what is comfortable,
For we are all apprentices,
In a craft where no one becomes the master,

Do not let pride,
Blind you from all above,
It's okay to spend time on your knees,
When there is no place else to go,

No matter the success,
No matter the awards you receive,

No matter how high a throne you may be seated,
You are still sitting on your own arse,
Stay humble,
For when you think you have all the answers,
Fate will remind you,
That you haven't learned much at all.

Humor 24 May 24

Humor,
Is a valuable part in the art of leadership,
It wakes up the brain,
Helping you overlook the unbecoming,
And tolerate the unpleasant,

A sense of humor,
Is deeper than rooted opinions,
It can have a positive influence,
On all you share it with,

Humor brings balance,
It stimulates you to see old things,
In new ways,
And to see yourself,
As others see you,

Humor is essential to perspective,
Understanding life in proportion,
It is more than laughing at things,
But in understanding them,
And understanding life,

Humor is healthy,
A needed armor in today's difficult times,
For laughter that comes from the heart,
Can heal most wounds,

Surviving today's unrestrained culture,
Requires a good sense of humor,
For it can silence your enemies in failure,
And humble you in moments of success,

Without a sense of humor,
You become blind to the world,
To human nature,
To the absurdity in our everyday life,
And if life doesn't make you laugh,
You didn't get the joke.

Patience 23 May 24

Patience is not the ability to wait,
But the ability to remain calm in chaos,
Constantly taking action,
To create positive opportunities,
And believe in yourself,
That you made the best decision available,

It is the willingness to stay where you are,
And live the situation to the fullest,
To listen,
In the belief that something hidden,
Will reveal your solution,

Do not search for the answers,
That cannot be given to you now,
Live the questions now,
You may learn more about yourself,
Then the situation you are facing,

Fear will keep you stuck,
Learn to conquer yourself,
When the storms of endurance,
Disturb your soul,

Patience is bitter, conquering,
Patience is power, a virtue,
But sometimes you must take the moment by the throat,
And act,

Patience achieves more than force,
And no one can influence your mind,
Unless,
You give them the power to do so,

Two of the hardest tests for leaders are,
The patience to wait for the right moment,
And the courage to act,
When everything around you tells you to stop,

Remember,
We are all made from one another,
And life waits on us,
A thousand time more,
Then we wait on life

Mistakes 22 May 24

To err is human,
But our culture states otherwise,
Zero tolerance is the norm, the law,
And no one is exempt,

Mistakes happen, expect them,
Perfection can be sought,
But know it doesn't exist,
For without imperfection,
None of us would exist,

Shortsighted decisions,
Come from inexperience,
Let them suffer the results,
At the same time,
Provide balance and support,
By not setting policies,
That magnify one mistake,

Expect mistakes,
But one should always make new ones,
Repeating the same mistakes,
Is a hallmark of a dim consciousness,

Mistakes expand perception,
To admit those mistakes,
Shows the ability to learn,
And growth personally and professionally,

The shallow,
Measure a person by their success,
Forgetting that failures,
Are the foundation of effective people and organizations,

The eyes above are always watching,
Scrutinizing your every action,
And if you are a leader,
One mistake is all they see,

Forgiveness,
Is the greatest power ever wielded,
People will rise above their faults,
For they are more than their errors,
And the perceptions branded upon them.

Intentions 21 May 24

There will always be disparity,
Between actions and intentions,
Between why,
And what the hell were you thinking,

There are intentions,
And actions of those intentions,
True intentions will make themselves known,
Whether evasive or productive,
They will harvest what you sow,

Misunderstandings,
Can distort the motives of pure intentions,
See the spirit behind the deed,
Beyond the action,
And find its true purpose,

Expectations are often hidden in silence,
Trapped by monuments never built,
Those who cannot look beyond past perceptions,
Will see a future with prejudiced eyes,

Look beyond the world of comforting illusions,
And deceiving realities,
See more than your feelings,
Use more than your experience,
To discover the truth,

Never judge someone on their intentions,
For accidents happen, and mistakes take place,
And sometimes facts,
Are nothing more than pathetic consequences,

Consider,
You may not always be right,
In a way you interpret other's words or actions,
Take a moment,
Disarm your assumptions and delusions,
For reality without knowledge,

Is weak and useless.

Complacency 16 May 24

So many are deceived by comfort,
And chose to wait,
They get used to waiting,
Until mediocrity becomes their success,
And failure,

Defective methods,
Deceive us into thinking,
That one way, is the only way,
And going through the motions,
Is truth,

Seek knowledge and wisdom,
And the understandings that vivify your existence,
For once your soul is awakened,
You will never return to a sleepwalking state of mind,

Do not view yesterday's achievements,
With the same pleasure as when it was completed,
Look forward with lessons learned,
For stagnant minds,
Are conditioned to fail,

Let today's victories,
Become rehearsals for tomorrow's conquests,
The beginnings of your journey,
Do not let the failures of others,
Become yours,

Life's greatest tragedies,
Are found in complacency,
So do not fear inspiration,
For a leader's legacy is bound by their excellence,
Not by fear,

The truth behind a successful leader,
They are always learning, moving forward,
Confident but not comfortable,
To do otherwise is to settle,

And leadership is only over,
When you become complacent.

Expectations 8 May 24

Will you interact with your world,
Play a part in the story,
Or do you prefer life to remain untitled,
A chapter without a heading,

Understandings,
Can become misunderstandings,
If there is no commitment,
No responsibilities assured,
No objectives set,
And no values shared,

Expectations,
Go both ways,
They expect you to know the answers,
Before knowing the questions,
And you expect them,
To have the same dedication as you,

Be ready for the storms,
If you're going to say what you want to say,
You're going to hear,
What you don't want to hear,

Good leadership,
Means making order of successes and failures,
Do not expect acceptance or commitment,
If you cannot show it in yourself,

Learn to let go of hidden expectations,
Those hopes and desires you have for yourself and others,
That you never communicate openly,

Separate your observations,
From your preconceptions,
See what is,
No what you expect,

You are your own worst enemy,

Don’t go swimming,
In other people's expectations,
Or assuredly,
You will drown in your own.

Talent 29 April 24

Talents are meant to be used,
They do not make you better,
Or worse than others,
A popular name will only carry you so far,
And mediocracy,
Knows nothing higher than itself,

Those that cannot see them,
Will raise suspicions,
For talented minds,
Are always targeted by mediocre ones,

Successful leaders,
Discover, cultivate, and learn,
The talents of those they lead,
They create opportunities for their potential,
And inspire them to live up to it,

Do not deny,
Any sense they might make,
Open your mind beyond your pride,
For great ideas,
Can emerge from useless fragments of thought,

Talents are a gift,
And what you make of them,
Is a gift in return,
The more you invest,
The greater the yield,

For every unique talent,
There are also unique needs,
When these needs are matched with opportunities,
You may discover abilities,
You never knew existed,

Work on your gifts,
And spend them all,

Do not wait for later,
For anything not used freely,
Will become lost,

Awaken the sleeping giant,
Have the courage to follow even in doubt,
For they can do, what few can do,
And they see, what even fewer can see.

Black Sheep 29 April 24

Black sheep are not lost,
They are the outliers,
The trailblazers,
With no reason to choose sides,
Because no side will choose them,

They choose to live authentically,
With the courage to make their own path,
They refuse to be what everyone else is,
And ignore the opinions of others,

They do not fit within the heard,
They may be the minority,
But often leave a greater impact,
Then the rest of the flock combined,

Those with slow seeing,
Fear standing on any edge,
But the black sheep are seekers of liberation,
And will challenge the status quo,

So many hide their true selves,
In attempts to fit in,
But the black sheep never will,
For you don't have to like them,
And assuredly they won't care,

The brotherhood of man,
Does not imply equality,
For all organizations have their fools and geniuses,
Black Sheep and saints,
Independent successes and worldly failures,

So don't try to fit in,
When you are born to stand out,
They hardest thing about being a black sheep,
Isn't the praise you won't receive,
But the differences you bring,
That no one else wants.

Paralyzed in Silence 14 SEP 23

Without wisdom or conviction,
They drive you to your knees,
Demand your ignorance,
And define your expectations,

They play upon your fears,
The ones created long ago,
By implementing safeguards to thinking,
So reality cannot be decoded,

The madness wears you down,
And you will agree,
So you will comply,
And listen without question,

They chip away at you,
So the pieces never fit,
And you no longer remember,
Who you were,
And what has been taken away,

They proclaim heroes are the stuff of fiction,
So you will acknowledge your own nothingness,
And surrender yourself,
To your own prison,

It is easy to lose faith,
And become lost in the silence,
The darkness may hide many things,
But it can never delete the truth,

You must listen through the silence,
The projected chaos forced upon the air,
Unlearn the inauthentic principles,
And take back,
That which has been given by God.

Emotional Intelligence 22 March 24

Fear does not exist,
Except in the mind,
Met your failures as learning experiences,
For they will become the core of your success,

Who you are,
Is a state of consciousness,
Changing with time and experience,
Adapting with every wave of emotion,
As life comes and goes,

Never react emotionally to criticism,
Analyze yourself first,
Determine if it is justified,
If it is, correct it,
If it isn't, move on,

Emotional intelligence,
Is knowing what energizes you,
And what drains you,
It is how you lead when calm,
And when stressed,

Setting healthy boundaries,
Requires the ability to say "NO",
And being willing and able,
To enforce the "NO" when needed,

Never speak out in anger,
Or act out of fear,
For we are at our most dangerous self,
When unconscious of our actions and behaviors,

Be clear about your expectations in life,
Master your emotions and reactions,
If you give someone the power to define you,
You have given them power to control you,

We cannot predict the future,
Or things that occur in the medley of life,
But you can decide what happens within,
How to take it, and what to do with it.

The Eternal Truth 22 March 24

Look at our society today,
Responsibility is being ignored,
Consequences are fading,
And identities lost,

So many are eager,
To take their obedient place in the machine,
Let the wheel spin without looking,
Accepting the official doctrine,
Without question or understanding,

But no one can turn you into a servant,
Unless you allow them,
No one can change the eternal truth,
Unless you allow them,

Ethics cannot be developed overnight,
Or by official declaration,
It is developed by years of training,
And performance of duty,

Choice is the root of all morality,
Without choice, values have no meaning,
Integrity without knowledge,
Is weak and useless,
And knowledge without truth,
Is dangerous and appalling,

Keep your own commandments,
In the greatest of solitude,
For no witness or audience,
Can hold you accountable to the level of yourself,

Live in accordance with your capabilities,
For the eternal truth begins with you,
You can never escape responsibility,
Or the rewards of your actions,
With duplicity and excuses.

Competence 19 March 24

Be patient,
There is no set time for skills to be obtained,
Each is acquired uniquely,
You must understand your priorities,
Before you can meet them,

Stretch yourself,
Beyond your perceived limitations,
For experts were once amateurs,
Who kept learning and refining their vision,

Keep your wits in a crisis,
Improve solutions when every second counts,
Don't be the driver,
Of the truck that runs you over,

You own your skills,
And the key to value creativity,
This is your capital,
You decide who, when,
And where to invest it,

Be aware of your own boundaries,
Seek enlightenment, but remember your origins,
You can fake virtue for an audience,
But your actions will always speak the truth,

Competence creates confidence,
Leaders will believe in your potential,
But your colleagues,
Will make you live up to it,

Check your leadership for leaks,
For moments of choice,
Can become moments of truth,
And the true test of competence,
Always lies in the end result.

Authenticity 19 March 24

You are the fabric,
That keeps your dreams alive,
Do not compromise something beautiful,
To create something fake,

Don't let others' opinions,
Affect your decisions,
Or others' ideas,
Limit who you are,
Live your reality,
Not theirs,

Authenticity is the privilege of a lifetime,
To become who you truly are,
Not the illusions,
Of an inflated reality,

Live an authentic life,
For your strength lies in the daily choices you make,
One after the other,
Creating external ripples,
Of a life well lived,

Never trade reality for safety,
Cast off your shadows,
For troubled thoughts,
Create thousands of distractions,

Authenticity is invaluable,
Embrace your imperfections,
Ego is the need to prove what you say,
Authenticity is living life in your own truth,
Not the actions of your past,

Explore new horizons,
Of what it means to be completely you,
Be selective with your time and energy,
For they are both limited.

Delegation 14 March 24

It is the ultimate sign of trust,
Between leaders and team members,
Empowering people to develop,
While taking ownership and responsibility,

Trust is powerful,
But is also fast and fleeting,
Trust is also reciprocal,
If you give trust,
It will be given back,

Delegation is not about getting things done,
Or passing off tasks,
But investing and evolving others,
To accomplish the mission,

Delegation works,
Providing the one delegating works too,
It requires willingness to pay for short term failures,
In order to gain long term competencies,

Leaders must select the right people,
To do what needs to be done,
And have enough self-restraint,
To keep from interfering while they do it,

Delegation passes on actions,
Not responsibility,
You may share it with others,
But your portion is not diminished,

Each delegated task,
Must be accomplishable and well defined,
Leaders should concern themselves with results,
Not the application of their activities,

Don't try to do everything yourself,
Because you can't,

Break down ideas,
So everyone can see the big picture,
You can always modify your plan,
But only if you have one.

Masters of the Spotlight 14 MAR 24

They have an uncanny ability,
To find an audience,
As outside validation,
Is their only source of nourishment,

Masters of the spotlight,
Performers with little talent,
Crippled souls,
Addicted to attention,

Empty vessels,
That create the most noise,
Skilled chefs,
That cook up unnecessary drama,

They try to fill themselves,
At the expense of others,
Even an act of kindness,
Is an opportunity to grab the attention,

Their sense of entitlement,
Becomes their culture,
They love the attention so much,
The strive to answer rhetorical questions,

They are champions,
Of making mountains out of sinkholes,
Superheroes of self-importance,
Who's cries for attention,
Are like fingernails across a chalkboard,

With no remorse for their victims,
They climb the slippery pole of validation,
Seeking the center of the universe,
So they are the star of their own show,

The seek the spotlight,
But avoid overexposure,

They must fill their daily quota,
Before the light exposes the truth.

Blue Falcon 13 March 24

Masters of disguise,
And bridges burnt,
They know the least,
But always say more,

They are experts in their field,
But only when the boss is watching,
Seeking the spotlight,
So others will not shine,

They arrive,
At the most inappropriate times,
Rising upon the shoulders of others,
Pointing fingers,
To hide their own deficiencies,

Addicted to attention,
They infiltrate the masses,
Twisting the truth,
So you only see,
What they want you to see,

Fawning parasites,
Who are fond of starting fights,
Never caring about the actions,
As long as it justifies their means,

Narcissistic disasters,
Never understanding what's going on around them,
And they love convenience,
For it is guaranteed results with zero risk,

They will judge you,
Shame you, condemn you,
Leave you at the most inopportune moments,
All with a teeth bearing smile,

But time is not their friend,
For it will expose who they truly are,

Champions of the friendless many,
Who talk behind your back,
As they leave you behind.

The GIF 8 March 24

They are known by many names,
Mythical creatures that manipulate the mind,
Changing standard procedures,
To fix that which is not broken,

They appear out of nowhere,
When you least expect it,
Bringing ideas after deadlines,
Leaving no time to execute,

The fly in with deviations,
Whims found floating between their ears,
Spreading their confetti of confusion,
And then vanish,
When it's time for action,

They are sinister goblins that give directions,
Without time or assets,
Manpower,
Or resources to accomplish the mission,

They are blue sky ninnies that change tactics,
In the middle of implementation,
Never informing those involved,
And never suffering the consequences,

They are unicorns of brainstorming,
Explaining the complexities of simple design,
Oblivious,
To the second and third order effects they bring,

Death by PowerPoint,
Is there weapon of choice,
When they weave their magic,
You know the best parts of your job,
Are about to be tossed into chaos,

The GIF will never be truly defeated,
For they lurk in places,

Where leaders are disconnected,
Preying upon those,
Who cannot think for themselves.

Iron Strings 7 MAR 24

Do not be satisfied with other's stories,
Unfold your own truth,
Be whole and complete,
What others say or do is peripheral,

Risk curiosity and wonder,
Reveal your human spirit,
For the moment you doubt you can,
You won't,

Take decisions,
And make them the right decision,
You may not change opinions,
But you can change destinations,

Inspire others,
But make them work for it,
There is life after mistakes,
And the voice of others,
Will never validate you within,

Confront your fears,
As if they were opportunities,
For handicappers will try to choke your potential,
And create impossible conditions,
To cage you within their lies,

When you believe in yourself,
You don't have to convince others,
Or seek their approval,
Trust in yourself,
And others will follow,

Feel your hearts vibration,
Trust in the iron strings that shake your life,
Don't hold back from the experience,
Control your own destiny,
Or someone else will.

Leadership 5 March 24

It is performing at your best,
When not at your best,
Believing passionately in everything you do,
Never compromising your beliefs,

It is understanding with kindness,
Compassion in errors,
Winning with honesty,
And commitment with veracity,

It is never crossing the boundaries of attitude,
You must choose your own,
For it is more important,
Then the events happening around you,

It is never knowing,
All that you know,
But never doubting,
How much you truly care,

It is changing others' perceptions,
Be delivering more than you promise,
Accessing the hidden places,
Where few dare to tread,

It is arguing facts with validity,
Calculating every word, every step you make,
Being creative when inventing ideas,
And discipline while implementing them,

It is not a label you give yourself,
But a picture,
Its essence focuses on people,
The mission,
And all their demands.

Cheap thinkers,
Will never consider your commitments,

For you have assumed a responsibility,
That few can ever master.

Full Circle 26 FEB 24

Life is not linear,
It has its ups and downs,
Narrowing at times,
And widening,
Until it joins the motion of infinity,

To come full circle,
A celebration of growth and freedom,
It is life in every breath,
And the choices you make within them,

To come full circle,
Is to see with innocent eyes,
A reflection of eternity,
Radiating from the mind, body and soul,

To come full circle,
Takes patience,
Perseverance to see the dream,
And that which you protect,
Is what you receive,

To come full circle,
Back to where you began,
A journey over the same ground,
As if it were the first time,

Don't forget the you,
Behind the story,
Hold on to those days,
When things discarded become priceless,
And those lost,
Become loved,

To come full circle,
A path where the ordinary,
Becomes the extraordinary,
Back unto your true self,
To where you wish to be,

Where you have been,
And always were.

The Emotional Truth 19 May 22

It’s hard to forget the pain,
When it has become your world,
And peace can never be found,
By numbing the pain with false tongues,

Life has thorns,
Wherever there is light,
There can be darkness,
You cannot avoid the burdens of daily living,
By seeking tattooed delusions of escape,

To understand the emotional truth,
You must quiet the mind,
You can’t keep drinking from the same cup,
And expect and empty glass to fill itself,

Can you look at the pain,
Without the voice inside drawing conclusions,
Searching for what you think you are owed,
Only to find your life is taxed,
With a forethought of grief,

Looking the other way,
Won't change the view,
It’s not enough to have won the war,
If you cannot create peace within your storm,

You deserve a life,
Of the smallest freedoms,
So make your peace,
And in the silence,
Find your truth,

The emotional truth,
If you are chained to it,
You will never find a quiet world,
Until you make peace,
With the one you are in.

Would You 20 FEB 24

Traditions and principles,
Are systematically being effaced,
The belief in freedom,
Is no longer the dream,

So many are committed to nothing,
Service only if convenient,
With no sense of duty,
Selflessness only to self,

Would you give something,
That cannot be measured or bought,
Would you stand up,
Brave the cause,
Face the dust, blood, sweat, and tears,
With no expectations in return,

Would you stand at the gate,
And face the enemies of liberty,
Adhere to a common creed,
To ensure freedom still shines true,

Would you do,
What others won't do,
What others only say they'll do,
And protect those who won't,

There are those who will never know triumph,
Or defeat,
Will never bide the trumpets call,
For their prejudice,
Have blinded their souls,

What if your country asked you,
Would you answer the call,
Would you sacrifice today,
Knowing tomorrow is not a given.

Loyalty 24 JAN 24

You will receive what you earn,
And in exact proportion of how you serve,
You will not have commitment,
Without a relationship,

Loyalty is not an act,
But a habit,
It comes by action,
Not discussion or expectation,

It isn't easy to put into words,
It is neither an emotion, nor opinion,
But a state of mind,
A reflection of respect,
And essential to the basic principles of life,

You must be willing,
To see the most negative parts of people,
And still stand with them,
For in your darkest moments,
They will brave the shadows,
And bring you into the light,

Ignore outside distractions,
And live in their moments,
Celebrate their successes,
And teach in their shortcomings,

It begins with you,
At your core,
It is a pledge of truth,
Modesty in success,
And fidelity in the face of failure,

The greatest award a leader can ever receive,
Is the loyalty of their team,
It is not in plaques or ribbons,
But who they become by it,
And the opportunity to do more.

Masters of the Garden 24 JAN 24

Life is more fragile than you think,
It's not about being perfect,
But the effort you give to achieve perfection,
Effort is how change occurs,

Find the patience to wait,
And the courage to accept disappointment without anger,
Forget mistakes often,
For you are a combined effort,
Of everyone you've ever known,

Change your expectations,
Appreciate the little things,
Find the miraculous in everyone,
And see the message within the mess,

Be mindful of stagnation,
The enforcement of arbitrary uniformity,
Keep your eyes open to your mercies,
Be visible in victories,
As well as failures,

Appreciation can make a day,
Even change a life,
Treat others in a way that leaves no regrets,
Making excellence in others,
Belong to all,

It is not given,
The difference you can make,
Empower and influence each day,
For empowerment allows intuitive flashes,
And influence inspires architects of their own opportunities,

If you are the master of the garden,
Don't forget to water and seed,
Look at each bulb as it unfolds,
If you only stroll through,
You are the follower,

Not the master.

For God and Country 19 OCT 23

So many remain unknown,
To the gracious hand that provides our peace,
That strengthens us,
With wisdom, virtue, and prosperity,

So many hearts without souls,
So many frenzied outbursts without substance,
So many are intoxicated with broken success,
That grace is no longer preserved,

The indissoluble bond is broken,
Reason and experience are forbidden,
Patriotism is claimed without service,
And morality only when favorable,

Those with deceitful minds,
Have forgotten the blessings bestowed,
Our foundations erased,
By the relentless pillars of enlightened depravity,

The embodied teachings are now illicit,
Righteousness and freedom are no longer binding,
As the narratives sold to the young,
Demand unequivocal acceptance,

Intoxicated with broken success,
Wealth and influence judge the future,
Our symbol is stained,
And our ideals slowly contaminated,

For God and Country I have trembled,
Battled monsters of others choosing,
Waited while justice sleeps,
But the light of truth,
Will not slumber forever.

The Rock 10 OCT 23

The rock is your foundation,
It supports the structure as a whole,
It provides balance for faith, love and honesty,
And it must remain unyielding,

You cannot build on the ruins of the past,
Or make it up as you go along,
You must be patient,
Charge the tides,
If you wish to change the storm,

Stand upon the rock,
And your fears will fall,
Dare to face the troubled waters,
And the mountains shall move,

Do not fear the winds and rain,
Or the limitations imposed by distractions,
The winds of change will come and go,
But the rock is forever unchanged,

The rock is fearless,
It has no limit to time or space,
There is no army, no fortress, or darkness,
That can stand against its light,

The rock is not limited by your circumstances,
You must put faith,
In the power of faith,
For it will guide you through all turmoil,
With illumination and forgiveness,

You must find the rock,
Your rock,
For if your balance is weak,
So shall your foundation be.

Choices 21 SEP 23

They tell you the story,
Alluring ruses with attractive tales,
Where fact and fiction blend,
And the story,
Takes on a life of its own,

Fear and hate is their message,
To divide all from hope and light,
Creating a prison of promises,
And ignore the voices within,

In their lair it breeds,
And the potentate language is learned,
Where the art of rhetoric,
Becomes the law of rule,

Desperate for validation,
Inequity becomes the mandate,
They whip up your worst passions,
And call upon your highest fears,

Elevated to the top,
By unconscious ignorance,
A place where power and insanity,
Become culture and tradition,

No longer inspired by truth,
They create analogies,
So you will forget the questions,
And entrust the direction of our culture,
To the vagaries of amateurs,

Built on immoral compromises,
They fertilize the gardens,
Without thought of generations to come,
Freedom of choice,
Becomes no choice at all.

True North 15 SEP 23

A compass,
Never tells you about the destination,
It points you in the right direction,
And helps you navigate,
The complexities of this world,

It shows you what can be,
In uncharted territory,
Discovering critical points,
That line up with reality,

The wheel of time is set,
But the direction is always changing,
One leg of the journey is fixed,
The other,
Like life, is always moving,

So many never worry about a compass,
Until they are lost,
And so many get lost,
In the desert of their own making,

Find your stars,
To remember the way home,
Just because someone else's compass points north,
Doesn't mean you have to follow,

Even when the adventure is calm,
Keep your compass close,
You may submit to the unknown,
But never the unknowable,

True north,
Molds and constructs the soul,
It brings order to life,
Guides our conduct,
Showing what needs to be done,
And what should be left undone.

Character to Destiny 7 SEP 24

It is not a matter of chance,
But of choice,
You must be strong enough to face the world,
Or be crushed by its fears,

Other's opinions are not your purpose,
And your history is not your horizon,
You must be a rebel with a cause,
And an angel in temperance,

The winds of change blow where they blow,
Sometimes with gust and fury,
And others on a gentle breeze,
But destiny does not make house calls,
You must go out and find it,

Your imperative is to follow it,
Accept it,
No matter where it leads,
For there are no secret doors,
That lead to your purpose,

Impose your own plan of action,
Confront darkness with the illumination of truth,
For your reputation is what others think of you,
And your character is what God knows of you,

Never forget,
Whether it be storm or fire, peace, or grace,
The sword of destiny has two edges,
And you are one of them,

Dream outside the lines,
And find your path,
The one unknown to many,
For your character,
Will lead you to your destiny.

Service 30 AUG 23

It cost nothing,
But it can produce unexpected results,
Forgetting to do so,
Can cost you everything,

When you say it,
Mean it,
Feel the importance behind the words,
And your people will too,

The first responsibility of a leader,
Is to define their expectations,
The last is to show appreciation,
And in between,
Remember, you are a servant to your people,

Service,
Is the call to your existence,
Remember they are part of we,
And only we can succeed together,

As a leader,
Your people are part of your story,
Show appreciation in their successes,
And hope in their failures,

Be aware of the value of praise,
Given at the right moment,
For it can be the key to unlock doors,
That hide unseen,

Do not seek gratefulness,
Or you will always find ingratitude,
And those who are too arrogant to show it,
Will always find themselves alone,
In a crowded room.

The Pinnacle of Arrogance 9 AUG 23

They listen to themselves,
And tell you what will be done,
They know less about more,
And nothing about everything,

Cretinous values,
Are transmitted by bizarre memes,
Reciting the echoes in their ear,
And the lies from the teleprompter,

To cure their addiction,
They whisper honeyed words to desperate minds,
And wrap themselves,
In cloaks of patriotism,

To gull the masses,
They proclaim their expertise,
And those experts who challenge their statements,
Become sinister enemies of the people,

Shepherds of peace,
Are hunted and labeled,
Subjects become their prey,
Until their authority goes unquestioned,

To control the content of what's produced,
They demand your allegiance and ignorance,
As the antidote of human experience,
Is long forgotten,

The pinnacle of arrogance,
The tyranny of tyrants,
Insisting you pay for what you don't want,
Don't need,
Exercising their power,
For the good of their victims.

The Essence of Freedom 8 AUG 23

It cannot be declared,
For freedom is given by God,
And liberty is earned,
Through practice and belief,

You must go back,
To move forward,
You must know its essence,
To change it,

Freedom does not rest,
It must be exercised daily,
It is the continued act of striving,
And defying external forces,

Life without freedom,
Is a body without a soul,
You either work for it,
Or accept others servitude,

The right to be heard,
Does not mean the voice speaks the truth,
Pattering life around others' opinions,
Only leads to tyranny,

Freedom must be defended,
And requires responsibility,
As conformity to blindness,
Is the growth of surrender,

The essence of freedom,
Is how you learn from history,
You will never be free from its ghosts,
If someone else decides your liberty.

Fear and Arrogance 19 May 23

Fear is not living afraid,
It is self-discipline,
Self-knowledge,
Proof you can meet life on any terms,

Arrogance is not greatness,
It is a confidence that never runs away,
When you're wrong, you face it,
When you're right, you say nothing,

Expose your fear,
You don't have to breathe their pollution,
Make fear an understanding,
See the light, search for the truth,

Arrogance allows you to attempt the impossible,
It is confidence in your voice,
Humility in your gestures,
And the glow in your gaze,

Talk to your heart,
Believe in your reality,
For once you believe,
You can risk curiosity, wonder,
And any experience that reveals the human soul,

Live your life with fear and arrogance,
Fear,
That you may miss opportunities,
That you are asleep in life,
Arrogance,
Always learning to know yourself,
And still open doors to mysteries.

Monsters 22 MAR 23

They filter the weak,
From the strong,
Providing a corridor,
For the steeling of souls,

Obliterated and broken,
They stray from the light,
Turning into the hunger,
That denies their own reflection,

Not all monsters look like monsters,
Some are so ordinary,
You won't recognize what they are,
Until it's too late,

Some seek to become them,
And seduce you with their enlightened ways,
They feed on your attention,
And feast on your fears,

Others create their own,
In the icy depths of their mind,
And then fear what they created,
For what it reveals about themselves,

With human faces,
And souls consumed by shadows,
They become functionaries ready to believe,
And act without questioning,

No one forces you to become a monster,
But if you feed on the fears of others,
And inspire worship without reason,
Then you will create your own demons,
And not just the ones that live in your head.

Wisdom 10 MAR 23

Those with fragmented souls,
Tell me my values are obsolete,
My beliefs are outdated,
And my cause is unjust,

To them I say,
Any ideology that fails to recognize,
A person's worth or dignity,
Is archaic,

A new way of thinking,
Has changed the conditions for living,
The stage now belongs to the enlightened,
Where selfishness is mandated for survival,
And simplistic cure-alls are forced as lasting solutions,

Altered viewpoints,
Alter values,
Causing culture to disappear,
With the push of a button,

Allusions of order,
Insensible to the privilege of living,
Are fettered upon external opinion,
And contracted sympathies,

You cannot destroy the past by denying it,
You cannot cancel truth,
By imposing silence,
Nor erase the honor of history,
By not telling the story,

I still cling to ancient concepts,
I believe in God, Family, and truth,
I won't stay young forever,
But with age comes wisdom,
And wisdom never fades.

The Golden handshake 10 FEB 23

The golden handshake,
Relying on deception and deceit,
Is often rewarded,
Though no one knows why,

With narrow focus,
They boast their credentials,
A contested awareness,
Without true understanding,

They rise to their level,
On the industry of others,
Owning other's ways,
As they perfect their fabrications,

The smaller the burden,
The greater the storm,
A sentence witnessed,
By their own frivolous attention,

Prone to omission,
With misguided intentions,
Compensating quality for quantity,
And reckless emotions,
For sound judgement,

The illusion of adequacy,
Is their blindness in reality,
As they ascend the mountain,
By the grace of the golden handshake.

Lions 27 JAN 23

Lions don't hunt mice,
For you cannot justify the investment,
If you fail, get up and try again,
If you fail and conform to failure,
You will always invest in failure,

Lions learn to inspire,
Learn more, dream more,
Do more,
And become more,

Lions grow other lions,
You can never do it all alone,
Success is measured by triumphs of the led,
Not the lion,

Lions lead from the front,
Leading by example, not words,
Defining their expectations,
For ownership starts from the bottom up,

Lions do not require permission,
For failure does not fix itself,
You can make things happen,
Watch things happen,
Or ask from the sidelines,
What happened,

When things go wrong,
Lions first look within,
For those that say they are not responsible,
Have chosen to be irresponsible,

Communication is not an illusion,
And feedback is a gift,
Unbeatable Prides tell the truth,
And lions listen.

The Truth about Lies 1 SEP 22

Just because it isn't a lie,
Doesn't mean it isn't deceptive,
Those who speak in potions,
Are true craftsmen of destruction,

They lie to themselves,
And listen,
They become what they hide,
And cannot distinguish the truth within,

Once becomes always,
Opinions become fact,
And when truth is replaced by silence,
Silence becomes the lie,

Words become slippery and vicious,
Beliefs become manipulated,
Blurring the lines between reality and fable,
Depending on the crystal through which you see,

Under the engine of authority,
You are conditioned to the conditions,
Subjected to prefabricated interpretations,
Editing the soul according to trends,

The world is drowning,
So examine your words well,
For once you place the crown upon your head,
The stain will remain forever,

The truth about lies,
They have no music inside to dance to,
They have no respect,
And without respect,
You cannot love.

Destination 28 JUL 22

If you wish to catch a butterfly,
You must leave your cocoon,
For what wills to grow,
Will grow,

Can you do something,
Or will you do something,
The first is a question of ability,
The second render's ability irrelevant by your passion,

Reality is an excuse,
Waiting to be overcome,
When will always be later,
When, it could be now,

Running in circles is only an illusion,
Let your consciousness never sleep,
For if you do not see,
You will never know,

Listen to the silent voices,
For your point lies upon the horizon,
If you only focus on the confinements behind,
They will become your horizon,
And your destination,

Enjoy the pleasure of your journey,
Negotiate the unknown,
Sometimes you must break things,
So you don't forget the address on your invitation,

You are the impossible waiting to happen,
The destiny within the flame,
You must fight the lie,
That any purpose is too big,
Or you're too small to achieve,

Little movements,

Only the author can know,
From a place of purity and truth,
The brutality of the road,
Will guide you where you're meant to be.

Cult of Ignorance 13 FEB 22

Destruction breeds on the horizon,
It speaks from a place that has been taken,
Were vicious inspirations of uninformed impressions,
Hide behind barriers of traditions,
That are long since gone,

They appoint themselves the guide,
And force their feeble beliefs upon the innocent,
Invading their homes, schools, and libraries,
By filling the eyes with myopic dreams,

They share their drivel noise,
Eager to feed panic and suspicion,
Leaving you at the mercy,
Of their self-centered devotion,

They move around in their crowd,
Trampling on the eyes of others,
Ignoring the cries of the resistant,
As they attempt to define what is light,
To the deaf and blind,

That which they manifest is before us,
Irrational thought and attendant behavior,
Intolerance and impatient arrogance,
Exposing you to their culture,
And the crush that enforces it,

A cult of ignorance,
In the illusion of knowledge,
Living with your eyes wide closed,
A state where truth does not exist,
If you choose to ignore it,

The world is in a different place now,
The vision comes,
But no one is listening,
And herein lies the tragedy of our age,

That man knows so little,
Of man.

References used in gathering material for this book:

U.S. Army FM 6-22 Developing Leaders, NOV, 22

U.S. Army FM 22-100 Military Leadership, JUL, 90

U.S. Army Leadership Handbook, JAN, 20

Upenn.edu/effective communication to effective leadership

Leadership Theories for Career Growth

Top 12 Leadership Certifications to Improve Your Leadership Skills

Organizational Development, Brown, 8th Edition, 2011

Strategic Management, Pearce & Robinson, 14th Edition, 2015

Employee Training & Development, Noe, 6th Edition, 2013

ABOUT THE AUTHOR

Michael Morton is a retired Army Infantry Soldier. He served over twenty-four years, including multiple combat tours in Iraq and Afghanistan. He is the author of works such as "Behind the Camouflage", "The Last War", "The Anabel Project" learning to live, love, and heal with PTSD, "Poetry of Life", and the "Return to Innocence" series.

Made in the USA
Monee, IL
04 November 2026

67383422R00085